LeapFrog SchoolHouse™
LeapMat™

Installing the Batteries

Use three AA alkaline batteries

- Access the batteries of the LeapMat™ by removing the battery cover with a Phillips or flat head screwdriver.
- Install batteries as shown in the diagram on the inside of the battery compartment.
- Replace the cover.

As the batteries run low on power, the speech of the unit may become garbled or repeat itself. It is necessary to replace the batteries at this time.

Push the blue "Start" button and adjust the volume switch to low or high.

The LeapMat™ has 6 modes of operation:

1. **Press a letter to hear its name** – In this mode, when a student presses a letter LeapMat says the letter's name.

2. **Find the letter** – In this mode, students are asked to press a letter. Incorrect answers receive a fun sound to notify the student the answer is incorrect. Correct answers are met with an audio reward sound.

3. **Press a letter to hear its sound** – In this mode when a student presses a letter it says its sound. Both long and short vowels and hard and soft consonants are provided.

4. **Find the letter that says ___** – In this mode, students are asked to find letters based on the sounds they make. For instance, it will say find the letter that says, "/b/ as in 'bed.'" Again, incorrect answers receive a fun sound to notify the student the answer is incorrect. Correct answers are met with an audio reward sound.

5. **Spelling CVC (Consonant/Vowel/Consonant) words** – In this mode, LeapMat will ask the student to spell one of 32 different CVC words.

6. **Create-a-Word** – In this mode, the student is asked to press three letters to make a word. Over 450 words have been programmed. If three letters are pressed that do not make a word, LeapMat simply blends the sounds together and says, "Great sounds, try again."

Automatic Shut-Off

The LeapFrog SchoolHouse™ LeapMat™ will automatically turn off after a few minutes of non-use. There is also an "off" position on the function dial to the far left. It is best to store the LeapMat in the "off" position.

©1999 LeapFrog SchoolHouse

The Activities for the objectives are as follows:

Activity Objective	Activity Number	Activity Name
Learn the alphabet	1	Ready, Set, Sing
	2	Ready, Set, Read
	3	Ready, Set, Chant
	4	Slap the Mat
	10	What's Next?
	11	Four in a Row
Upper- and lower-case letter recognition	2	Ready, Set, Read
	4	Slap the Mat
	5	Hide and Seek
	6	Name That Letter!
	7	Leap's Big, Big Bag
	9	What's Your Letter?
	10	What's Next?
Distinguish initial consonant sounds	6	Name That Letter!
	7	Leap's Big, Big Bag
	8	Fill Leap's Big, Big Bag
	13	End of the Line
	16	Frogs on the Pond
	24	Leap on First
Recognize letters in sequence	9	What's Your Letter?
	10	What's Next?
Identify letter names and sounds	5	Hide and Seek
	11	Four in a Row
	14	Leap's Secret Letters
	15	LeapFrog Bingo
	16	Frogs on the Pond
	18	End Zone
	20	Word Bingo
Match letter names and sounds	4	Slap the Mat
	12	Zap! It's a Match
Distinguish initial, medial and final sounds	17	Name That Sound
Distinguish final consonant sound	25	Hop to the End
Make three letter words by adding the final sound	18	End Zone
Make three letter words using word families	19	All in the Family
Word identification	20	Word Bingo
Make three letter words	21	Consonant Cover-Up
Read and spell word families	22	Leap's Rhymes
Read and make new words with same initial consonant sound	24	Leap on First
Read and make new words with the same final consonant sounds as initial sounds	25	Hop to the End
Unscramble letters to make words	26	Leap's Mystery Word
Read, write and spell CVC words	27	Leap's Alphabet Words
Distinguish medial sounds	28	A Vowel Makes the Word
Printing letters	9	What's Your Letter?
	14	Leap's Secret Letters

©1999 LeapFrog SchoolHouse

Contents

Materials Provided

- *LeapFrog Phonics Mat*
- *52 Alphabet cards*
- *Bingo Boards–8 word and 8 letter boards*
- *Game Markers—15 each of 5 colors*
- *Pair of dice*
- *32 picture cards*

© 1999 LeapFrog SchoolHouse

Dear Teachers,

Learning phonics can be fun with the LeapFrog SchoolHouse™ LeapMat™. The LeapMat encourages students to participate in learning the alphabet, upper and lowercase letters, letter names, letter sounds, and simple words.

As reading specialists, we know how important it is for students to learn the connection between sounds and letters—a vital concept necessary for developing reading and writing skills. Spoken words consist of patterns of sounds. Written words consist of letters which represent the sound of spoken language. Beginning readers and writers need to master these fundamental skills:

- the alphabetic principle
- learn the name and usual sound for each letter
- perceive how sounds blend together in words from left to right
- sequence letters to form words
- hear words pronounced
- enjoy the learning process and extend this knowledge to reading and writing

The LeapMat activities in this guide present an introduction to the alphabetic principle and the basic phonics concepts in a format that is highly engaging and motivating for young children.

Each activity lists objectives, materials, preparation if any, and directions. The blackline reproducibles in the Appendix and the other materials included provide all the materials needed to play. Additionally, we have provided in the Appendix duplicates of certain materials including Alphabet Cards, Bingo Boards, and Picture Cards. The Matrix provides a visual log to track who has participated in each activity.

This guide provides 28 activities for interactive phonics learning. We have used these activities with our kindergarten, first and second graders, and second language learners with great success. Many activities include variations that provide different levels of play. These learning activities work well for students in any type of reading program.

The LeapMat activities effortlessly focus and maintains students' attention on the task of using letters and sounds in educationally productive ways. Students enjoy taking what they have learned in more traditional formats and transferring that knowledge to these activities. Because of the immediate feedback available, students are highly motivated and eager to learn and repeat favorite activities.

© 1999 LeapFrog SchoolHouse

Use of these activities in conjunction with the LeapMat provides students with word study and spelling development by encouraging an awareness of spelling patterns and word structure. They are designed to teach and reinforce alphabetic knowledge, letter/sound recognition, and reading and spelling of simple CVC (consonant, vowel, consonant) words based on common word families. These activities strengthen the skills and strategies critical in students' early literacy development.

The sequence of activities is flexible and can be tailored to individual classrooms. They are organized in order of students' literacy development. The first four activities, "Ready, Set, Sing," "Ready, Set, Read," "Ready, Set, Chant," and "Slap the Mat" are warm-ups that can be used at any time before beginning one of the other activities.

The objectives for each activity define the skills being developed. Later activities build upon skills introduced in earlier ones. In every classroom there is a wide range of literacy development. You as the teacher are the best judge of your students' readiness to participate in the different levels of learning. In our experience, often students who are having difficulty with initial reading need more practice and encouragement.

These learning games work best with groups of four or in teams of two. We also suggest that in certain games you choose a fifth player to be a Letter Selector. Ideally, this can be a student who needs extra practice. The activities can be modified to fit the circumstances and curriculum of any classroom or home learning environment. You can reinforce knowledge of particular letters by limiting an activity to only certain letters. To help students quickly grasp the idea, model or demonstrate the activity before beginning. Supervision helps students remain focused.

Cross-Age Buddy Reading in elementary schools is a powerful activity for student learning. These activities can be used very successfully as part of a Buddy Reading Program or a program of cross-age tutoring. Older students, Buddies, can be activity monitors and participants.

The LeapMat provides for two different types of assessment. Game options 2, 4, and 6 allow assessment of students' growing knowledge of letter names, letter/sound relationships, and independent spelling of CVC words. A second type of assessment is teacher observation during game playing. This assessment can be done as you observe and follow the students' early literacy development.

As the classroom teacher, it is important to become familiar with the six different modes of play:

1. **Press a letter to hear its name** – In this mode, when a student presses a letter, LeapMat says the letter's name.

© 1999 LeapFrog SchoolHouse

2. **Find the letter** – In this mode, students are asked to press a letter. Incorrect answers receive a fun sound to notify the student the answer is incorrect. Correct answers are met with an audio reward sound.
3. **Press a letter to hear its sound** – In this mode, when a student presses a letter, the LeapMat says its sound. Both long and short vowels and hard and soft consonants are provided.
4. **Find the letter that says ___** – In this mode, students are asked to find letters based on the sounds they make. For instance, the LeapMat will say "Find the letter that says 'ba' as in 'bed'." Again, incorrect answers receive a fun sound to notify the student the answer is incorrect. Correct answers are met with an audio reward sound.
5. **Spelling CVC words** – In this mode, the LeapMat will ask the student to spell one of 32 different CVC words.
6. **Create-a-Word** – In this mode, the student is asked to press three letters to make a word. Over 450 words have been programmed. If three letters are pressed that don't make a word, the LeapMat simply blends the sounds together and says, "Great sounds, try again."

In addition, you need to become familiar with the various learning games and how the interactive electronic alphabet mat works during a game. For example, during a game played on mode two, the letter will be prompted three times with an approximate ten second delay between prompts. After the third prompt the LeapMat will grow "quiet" until either the correct letter is pushed or the Help button is selected. If a correct letter is pressed the LeapMat will reply, "Awesome," "You're rolling now," or "Yes!" An incorrect response will be answered by, "Try Again" or a noise.

The LeapMat can be placed on the floor, on a table, or hung on a wall. Group play seems to work best on the floor with students facing the LeapMat. We have found it is useful to let students roll dice to determine the order of play, since every child wants to be first. Many of the activities can be played as competitive games. Several other variations are often more successful. For example, many activities are group-oriented, so play ends when the group has completed a task. Many activities focus on cooperative game playing.

Encourage students to share their own ideas for other phonics activities and games with the LeapMat. We have enjoyed participating in this project and look forward to hearing about your experiences.

Frederica Buel–Breuer   Janet Klisura–Fagre

Frederica Buel Breuer and Janet Klisura-Fagre are both reading specialists with the Piedmont Unified School District in California. Combined, they have over forty years experience as reading specialists, classroom teachers, and teaching graduate students. Janet has an Ed.D. in Language and Literacy Development from UC Berkeley. Frederica has an MA in Language and Literacy Development from UC Berkeley, and is in her third year as a Mentor Teacher in reading and children's literature.

© 1999 LeapFrog SchoolHouse

Activity #1 — Ready, Set, Sing

Objectives:
- Learn the alphabet in sequence
- Warm-up for activities using the LeapMat

Materials:
LeapMat
Alphabet Cards

Preparation:
None

Players:
2–4

Game:
Sing or chant the Alphabet Song with the students. Point to each letter on the LeapMat as you name it in the song.

Sing song again. Hold up each Alphabet Card letter and place it on the LeapMat. Have students in turn pick up the letters from the LeapMat.

Variations:
Have students sing or chant the song in different ways: quickly, slowly; ask them for their ideas; consider different grouping combinations of the song.

Let one or two students take turns accompanying the song with percussion instruments like a triangle or small drum.

© 1999 LeapFrog SchoolHouse

Activity #2 — Ready, Set, Read

Objectives:
- Reinforcement of alphabet sequence
- Upper and lowercase letter recognition
- Warm-up for activities using the LeapMat

Materials:
LeapMat
Alphabet Cards

Preparation:
Select an alphabet book (see list below)

Players:
2–4

Game:
Read an alphabet book to the class. As you read each page, point to the appropriate letter on the LeapMat. Ask students to chant the name of the letters with you.

Randomly pass out Alphabet Cards and have students in turn place cards on the LeapMat.

Suggested titles:

A, My Name is Alice by Jane Bayer
A Garden Alphabet by Isabel Wilner
Alphabatics by Suse MacDonald
Alphabet Soup by Kate Banks
Chicka Chicka Boom Boom by Bill Martin, Jr. and John Archambault
Have You Ever Seen . . . ? An ABC Book by Beau Gardner
Mr. Little's Noisy ABC by Richard Fowler
Potluck by Anne Shelby
Where Is Everybody? An Animal Alphabet by Eve Merriam

Variations:
Listen to alphabet tapes such as Chicka Chicka Boom Boom or a tape your class has recorded. Chant along with tape while pointing to appropriate letters on the LeapMat.

Pass out Alphabet Cards. Chant along with the tape. Have each child hold up the appropriate card and place it on the LeapMat.

© 1999 LeapFrog SchoolHouse

Objectives:
- Reinforce alphabet sequence
- Warm-up for activities using the LeapMat

Materials:
LeapMat
Alphabet Cards

Preparation:
None

Players:
2–4 or two teams of 2

Game:
Chant the alphabet using students' names: A is for Amy and Aaron, B is for Bryce, C is for Carlos, etc. If no student in the class has a name beginning with a specific letter, ask them to think of a name beginning with that letter.

Have each student find the beginning letter for his or her name from the alphabet cards. Then have each student find and press his or her letter on the LeapMat and place the alphabet card on the appropriate letter.

Variation:
As you chant, have students line up in alphabetical order holding the alphabet card representing their name. Have each student find, press and place the letter on the LeapMat.

© 1999 LeapFrog SchoolHouse

Slap the Mat

Objectives:
- Identify upper and lowercase letters and letter-sound matching
- Learn alphabetic order

Materials:
LeapMat
Alphabet Cards
Dice

Preparation:
None

Players:
2–4

Games:

Use mode 1. You can decide to use only lower case, upper case or a combination of letters for play. Deal four alphabet cards to each student, face up. Place the remaining cards in a pile for later use. Roll the dice to see who goes first.

The first player selects one card and places it in the appropriate place on the LeapMat, then presses the letter to hear its name. The student's partner or other members of the group can help the student if he or she cannot find the matching letter.

Player then selects another card from the pile and the next student takes a turn. Play continues until all cards are correctly placed on the LeapMat.

Variations:

Use mode 3 and play for initial sounds to develop letter/sound matching skills.

Deal all cards, face-up. The player who has the A card begins. Play continues in alphabetical order until all cards are played.

© 1999 LeapFrog SchoolHouse

Hide and Seek

Objective:
• Letter recognition and understanding that every letter has a name

Materials:
LeapMat
Game Markers
Dice

Preparation:
None

Players:
2–4 plus Letter Selector

Game:

Use mode 1. Begin play by rolling dice to see who goes first. Ask students to sit facing the LeapMat with their eyes closed while the Letter Selector presses any letter on the LeapMat. When the LeapMat names the letter, students open their eyes and the first player locates that letter on the LeapMat. The student repeats the name of the letter, presses the letter on the LeapMat and covers it with a Game Marker. She or he may select the next letter. Then the students close their eyes and repeat the process. As always, students should be encouraged to help one another.

Once a letter is covered with a Game Marker, it should not be selected again during the game. Play continues until all letters are covered with Game Markers.

Students can reinforce alphabetical sequence as they help put away the materials. Let the Letter Selector pick up the Game Markers in alphabetical order while the others chant the alphabet.

Variation:

Use mode 3 to play Hide and Seek.

© 1999 LeapFrog SchoolHouse

Objectives:
- Develop letter recognition
- Distinguish initial consonant sounds

Materials:
LeapMat
Game Markers
Dice

Preparation:
None

Players:
2–4

Game:
  **Use mode 2 or 4.** Roll the dice to see who goes first. The first player listens for the letter name or sound that the LeapMat says and presses the correct letter on the LeapMat. If players are unsure of the letter they can press the Help button and try again.

Other students may help the player if he or she cannot find the correct letter. Players receive a Game Marker for each correct letter or sound they identify. At the end of the game, players can total the number of Game Markers the group has accumulated.

Variations:
When the letter is found, the other players name a word that begins with that letter. If they cannot think of a word, encourage students to look for something in the room that begins with that letter; a picture from an alphabet book, a word from the word wall, etc.

Challenge version: Have players name words that end with the specific letters or letter sounds.

© 1999 LeapFrog SchoolHouse

Activity #7 — Leap's Big, Big Bag

Objectives:
- Upper and lowercase letter recognition
- Distinguish initial consonant sounds

Materials:
LeapMat
Leap's Big, Big Bag (See Appendix, pp. 43–46)
Picture cards for items in Leap's Bag: box, tag, rag, lid, and pot
Dice
Variations: Blank paper and pencils
 Set of 32 picture cards

Preparation:
Make copies of Leap's Big, Big Bag for each player. Find picture cards
for 5 items in Leap's Bag from the set of 32 picture cards.

Players:
2–4

Game:
Read Leap's Big, Big Bag to the class.

[a] Use mode 1. Show students the picture cards for each item in Leap's Big Bag,
one at a time. Have students take turns naming the picture and pressing the
letter on the LeapMat that matches the initial sound of the word.

Follow this procedure using the other 27 picture cards.

Variations:
Have students brainstorm for ideas of other items that could go in the bag. You
and the students can draw your own Leap's Big, Big Bag picture cards and
write the beginning sound of your object.

After you show your card and press the beginning letter, ask the first player to
say the name of the object you drew and press the letter for the beginning
sound. Then have the student name the object he or she drew and press the
beginning letter of that object.

© 1999 LeapFrog SchoolHouse

Activity #8 — Fill Leap's Big, Big Bag

Objective:
• Distinguish initial letter sounds

Materials:
LeapMat
Alphabet Cards
Leap's Big, Big Bag (See Appendix, pp. 43–46)
Game Markers

Preparation:
Make copies of Leap's Big, Big Bag for each player

Players:
2–4

Game:
Read Leap's Big, Big Bag. In this more challenging version of Activity 7, players pretend to pack Leap's bag by putting one item at a time into the bag and repeating in sequence what previous students have placed in the bag (just like in the game I Packed My Grandmother's Bag). Before starting, tell students which letters they will be working with for the game. For example, tell students they will be using only the letters in the top row or any six letters of your choice. Then place the appropriate Alphabet Card over these letters on the board as reminders for the students as they play the game.

 **Use mode 3.** Students take turns chanting, "I packed Leap's Big, Big Bag with ______." (Fill in the blank by repeating what others have put in the bag, followed by the name of their item.) When the student names his or her item they will press the beginning letter of that item, and place a Game Marker on the LeapMat. If a player can't remember the sequence of items, the other players can help out and play continues. Once a letter has been covered, no other items beginning with that letter can be placed in the bag. Play continues until all the letters in the row or the letters chosen by the teacher are covered by Game Markers.

Variations:
Fill Leap's Bag with items in alphabetical order.

Fill Leap's Bag using only items that begin with one specific letter.

© 1999 LeapFrog SchoolHouse

What's Your Letter?

Objectives:
- Upper and lowercase letter recognition
- Printing letters
- Letter sequencing

Materials:
LeapMat
Alphabet Cards (See Appendix, pp. 31–38)
Game Markers
Paper
Pencils

Preparation:
None

Players:
Small group and Letter Selector

Game:

Use mode 1. Ask players to look at the first row of letters on the LeapMat. Cover the chosen letters with the appropriate alphabet cards. Ask players to print one letter from the choices on their paper. Have them keep their letters a secret.

Letter Selector presses any letter in the top row. When the letter the LeapMat calls matches the one on a player's paper, the student calls out the name of that letter and places a Game Marker on the LeapMat. Letter Selector continues pressing letters in the top row until all students have matched the letter they wrote. Play continues in the same manner for the other three rows.

Variations:
Teacher selects letters to be used and marks them on the LeapMat by covering them with the appropriate alphabet cards.

Use mode 3. Have students match their letters with the letter sounds.

© 1999 LeapFrog SchoolHouse

Activity #10 — What's Next?

Objectives:
• Upper and lowercase letter recognition
• Reinforcement of alphabetical sequencing

Materials:
LeapMat
Game Markers
Dice

Preparation:
None

Players:
Two teams of 2

Game:
Use mode 1. Have one team (selected by roll of dice—higher number turns away) turn away from the LeapMat and close their eyes. The other team presses one letter on the LeapMat. Ask the team facing away to turn to the LeapMat and press the same letter, then the next letter of the alphabet in sequence. If the first letter pressed was S, the player would press S and T. Cover both letters with Game Markers so they cannot be played again.

Variation:
To increase difficulty level, have the first team press two letters in sequence. The second team presses the next two letters, in order (or the two letters preceding those already pressed).

© 1999 LeapFrog SchoolHouse

Activity #11 — Four in a Row

Objectives:
- Letter name and sound identification
- Review alphabetical sequence

Materials:
LeapMat
A variety of Game Markers, several of each color

Preparation:
None

Players:
2–4 or two teams of 2

Game:

Use mode 2. Give each player or team a set of same-colored Game Markers. When the LeapMat names a letter sound, players take turns finding the corresponding letter and placing a Game Marker on that letter. If a letter is repeated, the player puts a Game Marker next to the previous Game Marker.

The first player or team to get four of the same colored Game Markers in a row diagonally or across the LeapMat wins.

Variations:
Play until all letters on the LeapMat are covered with Game Markers. Each player or team can add up the number of Game Markers they have on the LeapMat at the end of the game.

Use mode 4 and play "Four in a Row" as described above.

© 1999 LeapFrog SchoolHouse

Activity #12 Zap! It's a Match

Objective:
• Match letter names and sounds

Materials:
LeapMat
Alphabet Cards (See Appendix, pp. 31–38)

Preparation:
None

Players:
2–4

Game:

Use mode 3. Have students take turns drawing an Alphabet Card. After saying the name of the letter and the sound it makes, the student verifies the answer by pressing the correct letter sound on the LeapMat. If correct, the student places it on the corresponding letter on the LeapMat. If not, the card goes back into the pile and play passes to the next student. Play continues until all the letters on the LeapMat are covered.

Variation:
Play as above except when a student or team correctly identifies a letter and sound, the student keeps the letter. If not, the letter goes back into the pile and play passes to the next student. At the end of the game, players/teams can count up the total number of letters they won.

© 1999 LeapFrog SchoolHouse

Activity #13   End of the Line

Objective:
• Identify initial consonant sounds in spoken words

Materials:
LeapMat
Alphabet Cards (See Appendix, pp. 31–38)
Dice

Preparation:
None

Players:
2–4

Game:

Use mode 3. Roll of dice determines order of play. Players take turns drawing an Alphabet Card. Ask them to say the sound the letter makes and name a word beginning with that sound: for example, / t / teeth. Students verify their answers by pressing the correct letter sound on the board.

If correct, the player places the Alphabet Card on the corresponding letter on the LeapMat. The winner is the one who completes a row on the LeapMat by placing the last letter in any row.

© 1999 LeapFrog SchoolHouse

Leap's Secret Letters

Objectives:
- Letter and sound recognition
- Printing letters

Materials:
LeapMat
Paper
Pencils
Variety of Game Markers, several of each color
Leap's Alphabet Board for Letter Selector (See Appendix, p. 95)

Preparation:
Make a copy of Leap's Alphabet Board for Letter Selector.

Players:
2–4 players and Letter Selector

Game:

Use mode 3. One student can be the Letter Selector for the game. Give several same-color Game Markers to each player using a different color for each student.

Ask students to print five letters on their papers and not show them to anyone. As Letter Selector randomly presses letters and marks them off on Leap's Alphabet Board, players check their "secret" list to see if any of their letters match. When a letter on a player's list matches the letter selected, he or she crosses it off. All players who match the letter place a Game Marker on the LeapMat for that letter. Play continues until all players have crossed off all the letters.

Variation:
Less challenging is to play mat row-by-row.

© 1999 LeapFrog SchoolHouse

Activity #15 LeapFrog Bingo

Objective:
• Letter name and sound identification

Materials:
LeapMat
Letter Bingo Boards, one per student or team
Game Markers

Preparation:
Make copies of Leap's Big, Big Bag for each player.

Players:
2–4 and Bingo Caller

Game:

Use mode 2. The Bingo Caller turns on the mat to mode 2. Players listen for the letter called out and check their Bingo cards for that letter. If their letter is called, players put a Game Marker on that letter on their Bingo Board. The Bingo Caller presses the letters called out by the LeapMat and puts the Game Markers on the letters on the LeapMat. This allows players to verify their letter.

The first player to cover a row of letters across, down, or diagonally calls out "LeapFrog!" Ask the winner to name the winning letters and check them against the Game Markers placed on the LeapMat. Play continues until every player gets a Bingo.

Variations:

Use mode 4 and play as described above. Players cover the letter on their Bingo Boards that match the sound of the letter the LeapMat says.

Using mode 2 or mode 4, vary the game by having students try to cover the four corner boxes, one specific row, one specific column or the entire board.

© 1999 LeapFrog SchoolHouse

Activity #16 Frogs on the Pond

Objectives:
- Letter names and sound identification
- Identify initial sounds
- Left to right, top to bottom sequence
- Counting 1–12

Materials:
LeapMat
A variety of colored Game Markers
Dice

Preparation:
None

Players:
2–4

Game:

Use mode 3. Give each player or team a different colored Game Marker. Students take turns rolling dice and moving the corresponding number of spaces around the LeapMat, starting at A and continuing through the alphabet sequentially, using the LeapMat as a game board.

When a player lands on a letter, the student presses the letter on the LeapMat, listens to the LeapMat say its sound, then names a word that starts with that sound. Other students may help the player think of a word. When players get to the end of a row they continue on at the beginning of the next row. Game ends when one player reaches Z.

Variation:

Use mode 1.

Note to Teacher: By using only one die, players will not go through the letters as quickly. You may also set a time limit to end the game so that when a player reaches Z, she or he returns to A.

© 1999 LeapFrog SchoolHouse

Activity #17 Name That Sound

Objective:
- Distinguish initial, medial, and final letter sounds in single syllable words

Materials:
LeapMat
Picture Cards (no words) (See Appendix, pp. 87–90)

Preparation:
Photocopy Picture Cards (no words), laminate if possible, cut apart.

Players:
2–4

Game:

Use mode 3. Review Picture Cards with students before playing so all players know the words the pictures represent, i.e., cat, not kitten, etc. Place picture cards face-down on a pile. Players take turns drawing one card, saying the name of the object, and pressing the letter for the beginning sound of the word. Players receive one Game Marker for each correct sound identified.

Variation:
Play for final consonant or medial vowel sound.

© 1999 LeapFrog SchoolHouse

Activity #18 — End Zone

Objective:
- Letter name and sound identification for early sounds

Materials:
LeapMat
End Zone Game Cards (See Appendix, pp. 60–68)
Pencils
Dice

Preparation:
Photocopy End Zone Game Cards and cut them apart.

Players:
2–4

Game:
Place the End Zone Game Cards face-down on a pile. Each player draws three End Zone Game Cards from the pile. Before beginning the game, allow players time to work out possible endings for their cards by writing on the cards in pencil.

 **Use mode 6.** Determine order of play through roll of dice. The first player shows one End Zone Game Card to the group. She or he presses the two corresponding letters on the LeapMat, then presses a third letter to make a word. If correct, the player takes a new card and places the used one face down. Play passes to the next student.

If a player cannot make a word or is incorrect, the End Zone Game Card is placed at the bottom of the pile. The student draws a new card to use later. Play continues until all cards are used. At the end of the game, players count up the cards they placed face-down.

© 1999 LeapFrog SchoolHouse

Activity #19 — All in the Family

Objective:
• Make three-letter words using Word Family Cards

Materials:
LeapMat
Word Family Cards (See Appendix, pp. 69–71)
Pencils
Game Markers
Dice

Preparation:
Photocopy the Word Family Cards and cut apart.

Players:
2–4

Game:
Place Word Family Cards face-down in a pile. Players select a card and think of three-letter words they can make. For example, for __ap they could make cap, gap, lap, map, nap, rap, sap, or tap. Players can write words on their cards.

Use mode 6. Roll dice to determine order of play. The first player makes as many words as possible by pressing a consonant and the two letters from the Word Family Card. Players receive one Game Marker for each correct word.

When a player cannot make any more words, he or she selects a new Word Family Card from the pile and the next player takes a turn.

Play continues until all the Word Family Cards have been used.

© 1999 LeapFrog SchoolHouse

Activity #20 Word Bingo

Objective:
• Letter name and word identification

Materials:
LeapMat
Word Bingo Boards (See Appendix, pp.72–79)
Game Markers
Blank Word Bingo Board for Bingo Caller (See Appendix, p. 80)

Preparation:
Photocopy several, 3–5, Blank Word Bingo Boards for Bingo Caller.

Players:
2–4 players and Bingo Caller (see note)

Game:

Use mode 5. Give each player a Word Bingo Board and Game Markers. As the LeapMat calls out the letters, allow players time to check their boards for the word. Bingo Caller presses the letters called out and writes the words in any order on the Blank Bingo Board. Then, he or she presses the correct letters on the LeapMat to make the correct word. If players have the matching word anywhere on their boards, they cover it with a Game Marker.

The first player to cover a row of words across, down, or diagonally calls out "LeapFrog!" Have the student read the winning words and check them on your list.

Variations:

Use mode 6. Have the player spell the three winning words by pressing the correct letters in order. The game continues with the same boards until everyone in the group has three words in a row.

Vary the game by having students cover only the four corner boxes, one specific row only, one specific column only, or the entire board.

Note to Teacher: The Bingo Caller's job is to keep a list of all the words called out. She or he may write the words on the Blank Bingo boards in any order. She or he will need several copies of the blank board to keep a complete list.

© 1999 LeapFrog SchoolHouse

Activity #21 — Consonant Cover-Up

Objective:
- Make three-letter words using CVC (Consonant, Vowel, Consonant) pattern

Materials:
LeapMat
Two sets of Alphabet Cards, consonants only (See Appendix, pp. 40–42)
Pencils
One set of Vowel Cards for each student (See Appendix, p. 39)
Two copies of CVC Master Page per student (See Appendix, p. 81)
List of CVC words (See Appendix, p. 82)

Preparation:
1. Photocopy two sets of the Alphabet Cards. Include Y as a consonant. Cut apart cards.
2. Photocopy one set of vowel cards for each student. Cut apart.
3. Photocopy two copies of the CVC Master Page for each student (and the optional list of CVC words).

Players:
2–4

Game:
Explain to the group the overview of the game:
    Step one: make a variety of CVC words using Consonant and Vowel Cards
    Step two: write CVC words on CVC master page
    Step three: make CVC words on the LeapMat using Consonant Cards only.

Model how to make a variety of words by dealing six consonant cards and vowel cards to yourself. Explain that CVC means words that follow the Consonant, Vowel, Consonant pattern. If you had N and P, you could, for example, make nap, nip, pin, pan, pen, or pun. Spell out the words on the LeapMat.

Give each player two copies of the CVC Master page. Deal six Consonant Cards and a set of Vowel Cards to each player. Place remaining cards face-down in a pile. Ask students to think about different CVC words they can make using their consonant and vowel cards. Give students time to write their words on their CVC pages.

 **Use mode 6.** The first player places two Consonant Cards on the corresponding letters on the LeapMat. The player then presses the beginning consonant, selects and presses a vowel, and then presses the final consonant to make a word. When a player makes a correct word, the Consonant Cards stay on the LeapMat. Play continues until all the consonants on the LeapMat are covered or all the consonant cards have been played.

Students, especially ESL students, who need extra help making words can refer to the CVC Word List.

© 1999 LeapFrog SchoolHouse

Activity #22 Leap's Rhymes

Objective:
- Read and spell common word families

Materials:
LeapMat
CVC Word Cards (See Appendix, pp. 83–86)
Paper
Pencils
Game Markers
Dice

Preparation:
Photocopy the CVC Word Cards. Laminate if possible and cut apart.

Players:
2–4

Game:

Use mode 6. Each player selects one CVC Word Card from the pile and writes as many rhyming words as possible on his or her paper. For example, a player with the card rug, could make bug, hug, jug, mug, tug.

Roll of dice determines order of play. Players take turns spelling all their rhyming words on the LeapMat. They receive one Game Marker for each correct word. If another player thinks of an additional rhyming word, he or she spells it on the LeapMat and receives a Game Marker. Play continues until either all cards are used or a preset time limit is reached.

© 1999 LeapFrog SchoolHouse

Three Makes a Family Word

Objective:
• Make new words using Word Family Cards

Materials:
LeapMat
Word Family Cards (See Appendix, pp. 69–71)
Game Markers, several of each color
Dice

Preparation:
Photocopy Word Family Cards. Laminate if possible, and cut out.

Players:
2–4

Game:

Use mode 6. Deal two Word Family Cards to each player.

Place the rest of the cards face down in a pile. Students take turns rolling dice and moving his or her Game Marker the corresponding number of spaces around the LeapMat, starting at A and continuing through the alphabet sequentially, using the LeapMat as a game board.

When a player lands on a letter, he or she tries to make a word using that letter in combination with one of the Word Family Cards. If the player has _ an and lands on P, the word would be pan. If the player cannot make a word using either card, play passes to the next player.

After each turn the player can discard one card and take a new one or keep both cards. Players should always have two cards in hand.

Note to Teacher: By using only one die, players will not go through the letters as quickly. You may also set a time limit to end the game so that when a player reaches Z, she or he returns to A.

© 1999 LeapFrog SchoolHouse

Activity #24 — Leap on First

Objectives:
- Identify initial consonant sounds
- Make additional words using the same initial consonant sounds.

Materials:
LeapMat
Game Markers
CVC Word Cards (See Appendix, pp. 83–86)
Paper

Preparation:
Photocopy the CVC Word Cards. Laminate if possible and cut apart.
Divide group into two teams.

Players:
2–4

Game:

Use mode 6. Teacher models game for students reminding them to listen for the first sound of the word. Roll of dice determines order of play.

Team 1 selects a word card from the pile, reads the word, and presses the letters to make the word on the LeapMat. Example: If the word is cat, they read it and press C-A-T on the LeapMat. Then they tell the other team to "Leap on First!" Game Monitor reminds team 2 to listen to the word and make a new word that begins with the same beginning sound.

Team 2 listens to the word and says a three-letter word that begins with the first letter of the first word. Example: If the first word is cat, the second team can make the word cap. They press the letters on the LeapMat to make that word. Play continues back and forth with players selecting word cards until time is up or until all the word cards are played. (Teacher determines time for play.) Team earns Game Markers for correctly made words.

Variation:
Play as described above, but without word cards. Students write down their own three-letter word list of three-letter word cards to use for play.

© 1999 LeapFrog SchoolHouse

Activity #25 — Hop to the End

Objectives:
- Distinguish final consonant sounds in single-syllable words
- Make new words using final consonants sounds as initial sounds

Materials:
LeapMat
Game Markers
CVC Word Cards (See Appendix, pp. 83–86)
Variation: Paper and pencil

Preparation:
Photocopy CVC Word Cards and cut apart.

Players:
2 teams with 2 partners or a group of 4 plus Game Monitor who
reminds teams to "Hop to the End."

Game:

Use mode 6. Teacher models game for students. Roll of dice determines order
of play.

Team 1 selects a word card from the pile, reads the word and presses the letters
to make the word on the LeapMat. Example: if the word is cat, they read it and
press C-A-T on the LeapMat. Then they tell the other team to "Hop to the
End." Game Monitor reminds team 2 to listen to the word and make a new
word that begins with the same ending sound.

Team 2 listens to the word and says a three-letter word that begins with the
last letter of the first word. Example: If the first word is cat, the second team
can make the word ten. They press the letters on the LeapMat to make that
word. Play continues back and forth with players selecting word cards until
time is up or until all the word cards are played. (Teacher determines time for
play.) Team earns Game Markers for correctly made words.

Variation:
Play as above but have players write down as many three-letter words as they
can make that begin with the ending sound. Game Markers are given for every
two words made correctly.

© 1999 LeapFrog SchoolHouse

Activity #26 — Leap's Mystery Words

Objective:
• Unscramble letters to make three-letter words

Materials:
LeapMat
Mystery Word Cards (See Appendix, pp. 91–94)
Pencils
Paper

Preparation:
Photocopy the Mystery Word Cards on paper and cut apart.

Players:
2 teams with 2 partners or a group of 4 plus Game Monitor

Game:

Use mode 6. Teams each select a Mystery Card from the pile of cards.
Example: a card with the letters I-D-L or a card with U-B-G. The object of the
game is to make a three-letter word from the three letters on the mystery card.
Players write the word on the card and press the correct letters on the LeapMat.

When ready, play begins and teams roll the dice to see who goes first to make
the words on the LeapMat. A Game Marker is given for each correct word.
Play continues until all of the cards are used.

Variation:
Have players make up their own three-letter mystery cards to use for the game.

© 1999 LeapFrog SchoolHouse

Activity #27 — Leap's Alphabet Words

Objectives:
- Read and write CVC words
- Spell CVC words correctly

Materials:
LeapMat
Leap's books and other simple CVC texts
Leap's Alphabet Board (See Appendix, p. 95)
Pencils
Game Markers

Preparation:
Make one copy of Leap's Alphabet Board for each player.

Players:
Whole class activity with 2–4 players at a time at the LeapMat

Game:
Give each player a copy of Leap's Alphabet Board and several books. Each player searches through books for CVC (Consonant, Vowel, Consonant) words and prints them in alphabetical order using the alphabet boxes. The goal is to find as many words as possible.

Use mode 6. When players finish their lists, they go to the LeapMat and take turns spelling their words by pressing the correct letters on the LeapMat. Players receive a Game Marker for each correctly spelled word.

Note to Teacher: Players can write their words on cards and play other LeapMat games which use CVC word cards.

© 1999 LeapFrog SchoolHouse

A Vowel Makes the Word

Objective:
• Distinguish the medial vowel in CVC words

Materials:
LeapMat
Picture Cards (See Appendix, pp. 47–50)
Pencils

Preparation:
Photocopy the Picture Cards.

Players:
2 teams with 2 partners or a group of 4

Game:

Use mode 6. Roll dice to determine order of play. Picture Word Cards are placed face down in pile. The first team draws a card from the pile and says the name of the picture on the card and checks the letters to make sure they correspond. Example: If the student says cat then she makes sure that the letters are C _ T. The team spells the word that goes with the picture on the card by pressing the letters on the LeapMat.

If the word is correctly spelled the team keeps the card and writes in the vowel. If not correct the card returns to the bottom of the pile. Play alternates between the two teams until all the cards have been played. Picture Word Cards are counted to see how many have been made.

Variation:
Individuals play instead of partners.

© 1999 LeapFrog SchoolHouse

Appendix

© 1999 LeapFrog SchoolHouse

D

H

C

G

B

F

A

E

© 1999 LeapFrog SchoolHouse

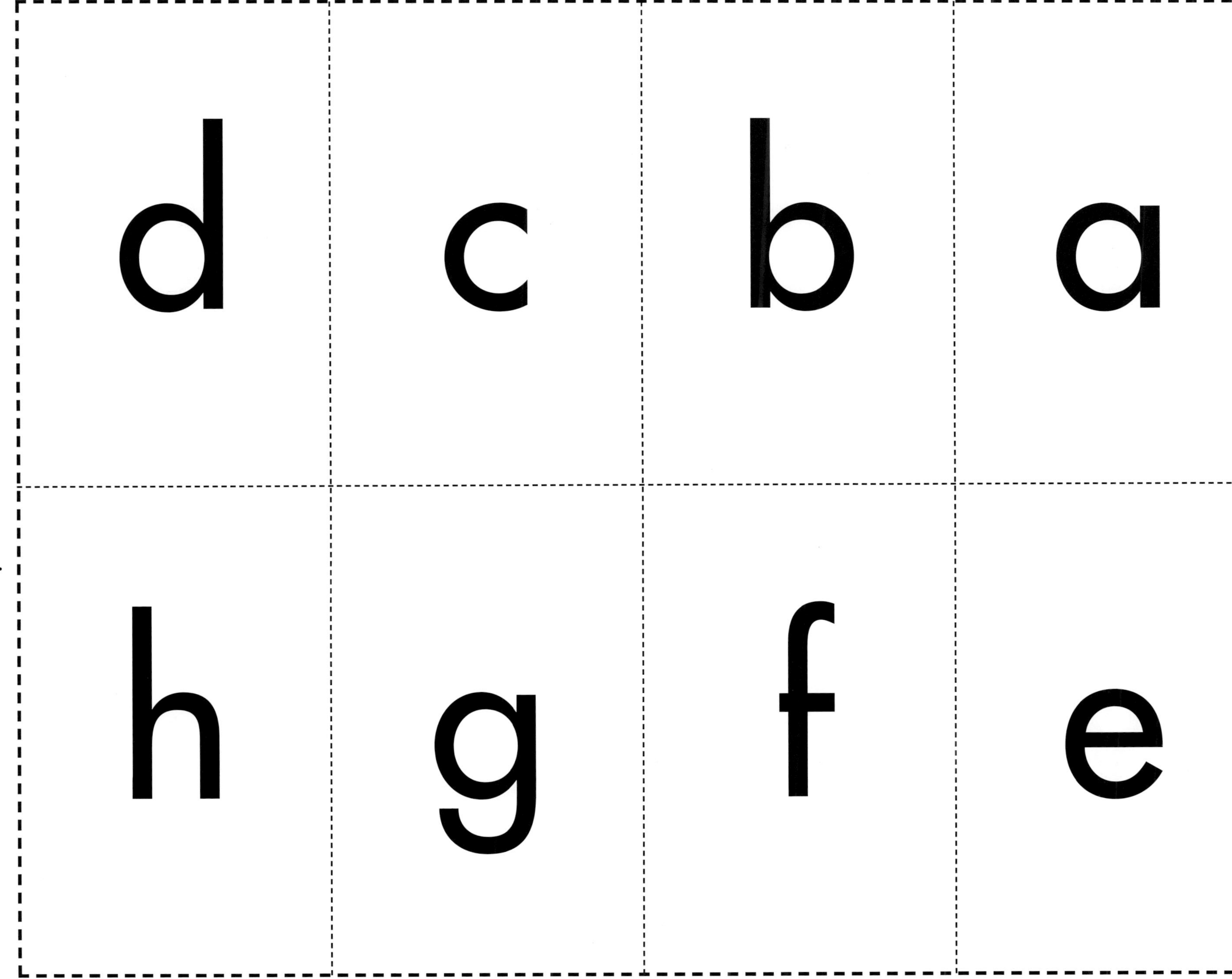

© 1999 LeapFrog SchoolHouse

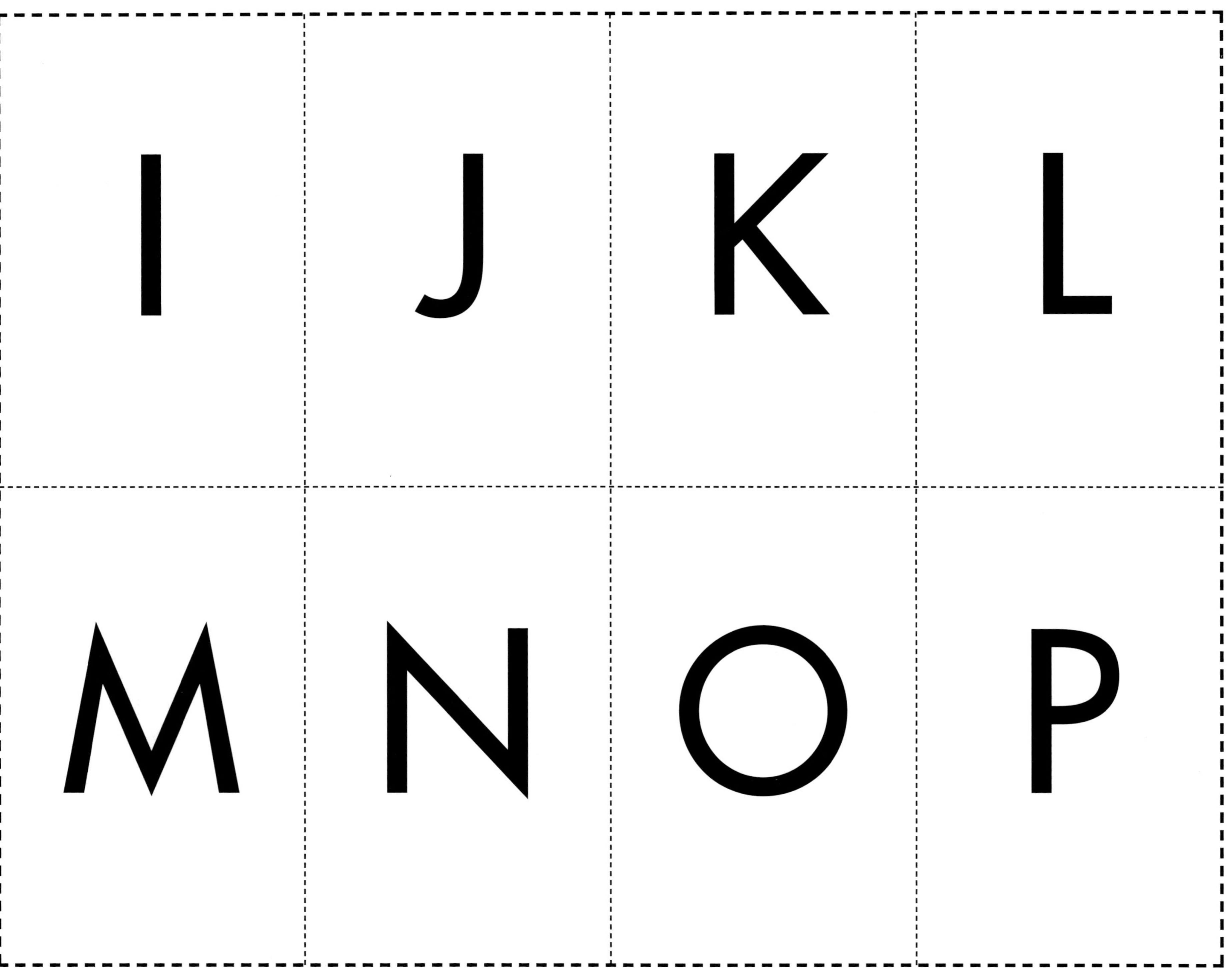

© 1999 LeapFrog SchoolHouse

© 1999 LeapFrog SchoolHouse

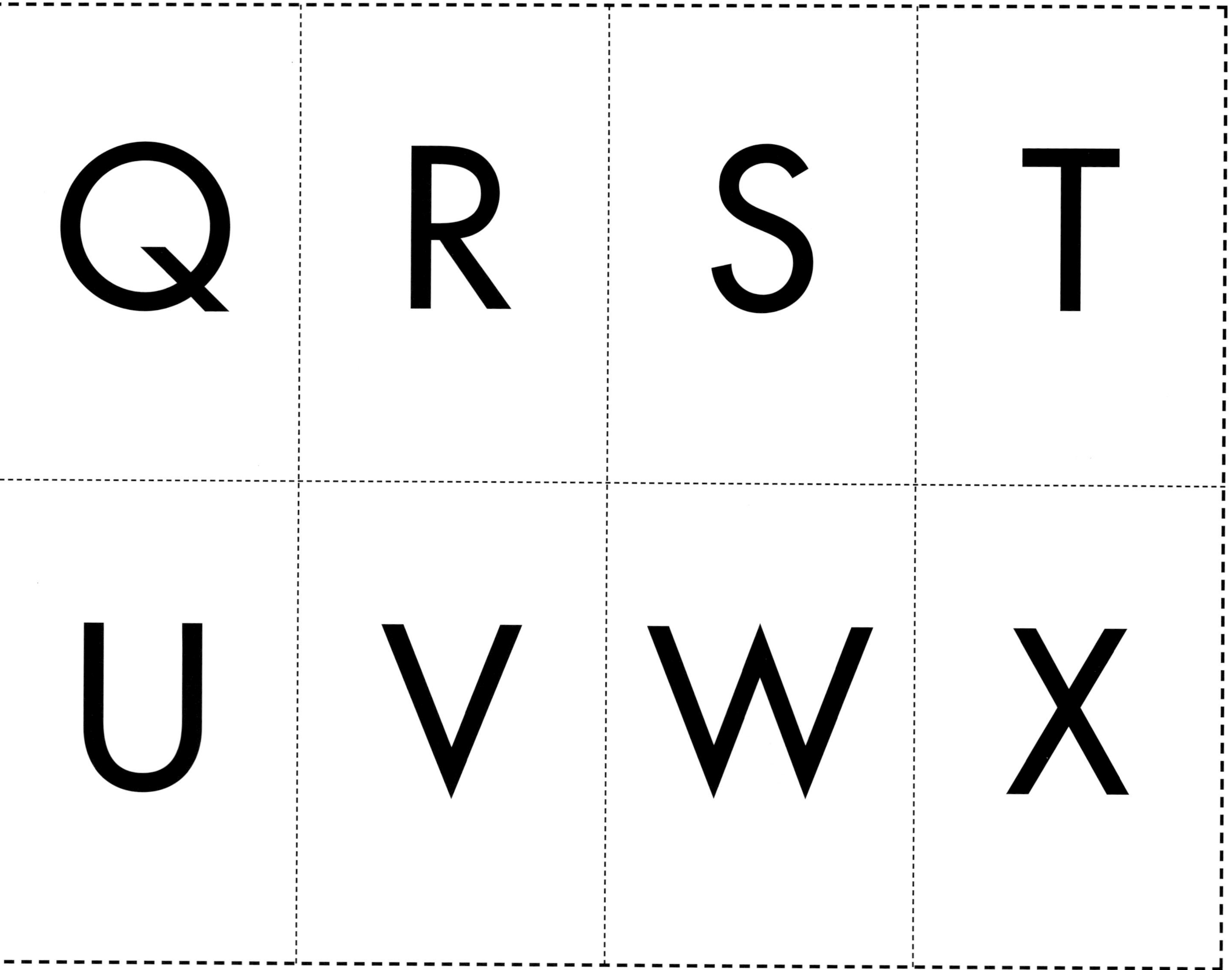

© 1999 LeapFrog SchoolHouse

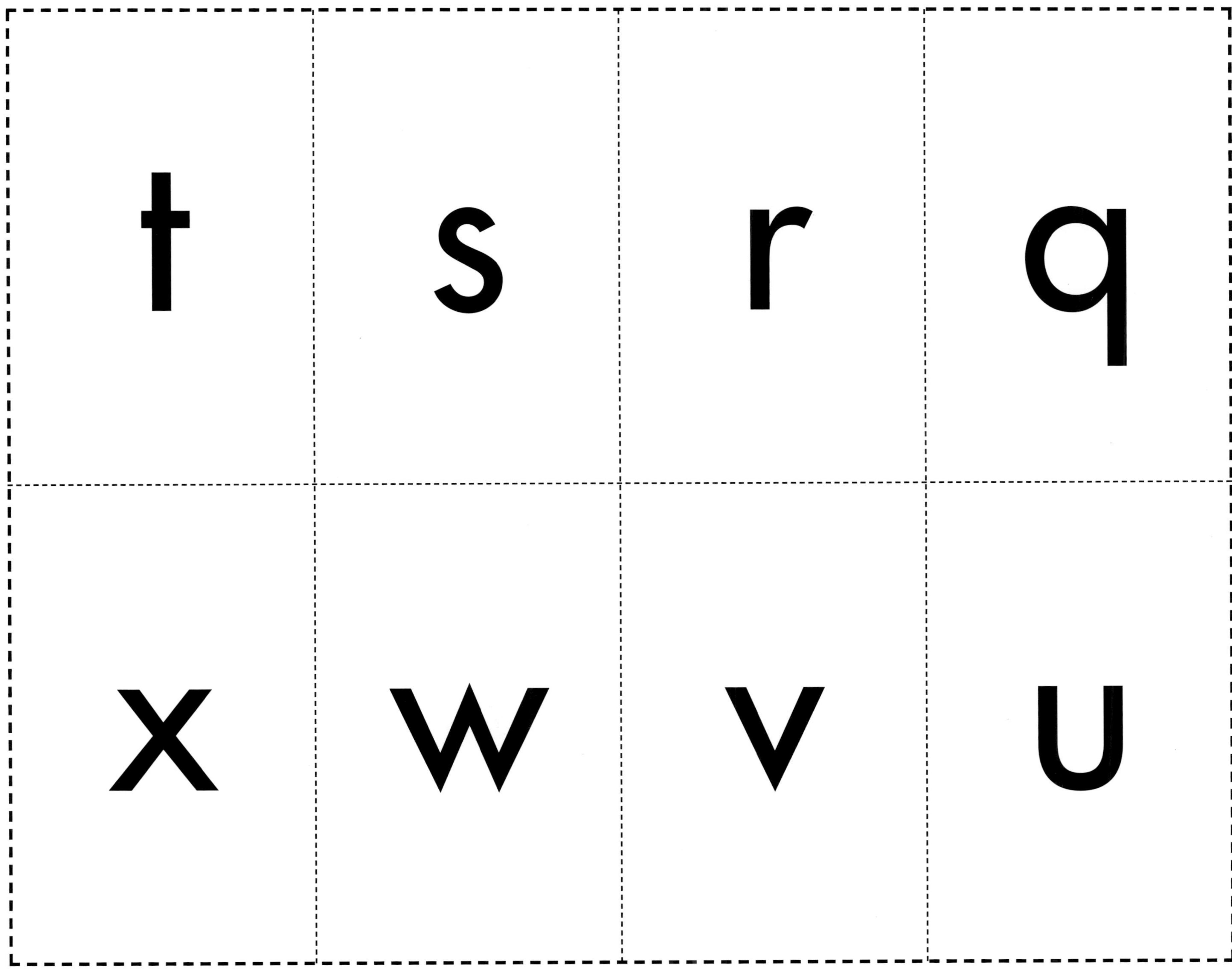

© 1999 LeapFrog SchoolHouse

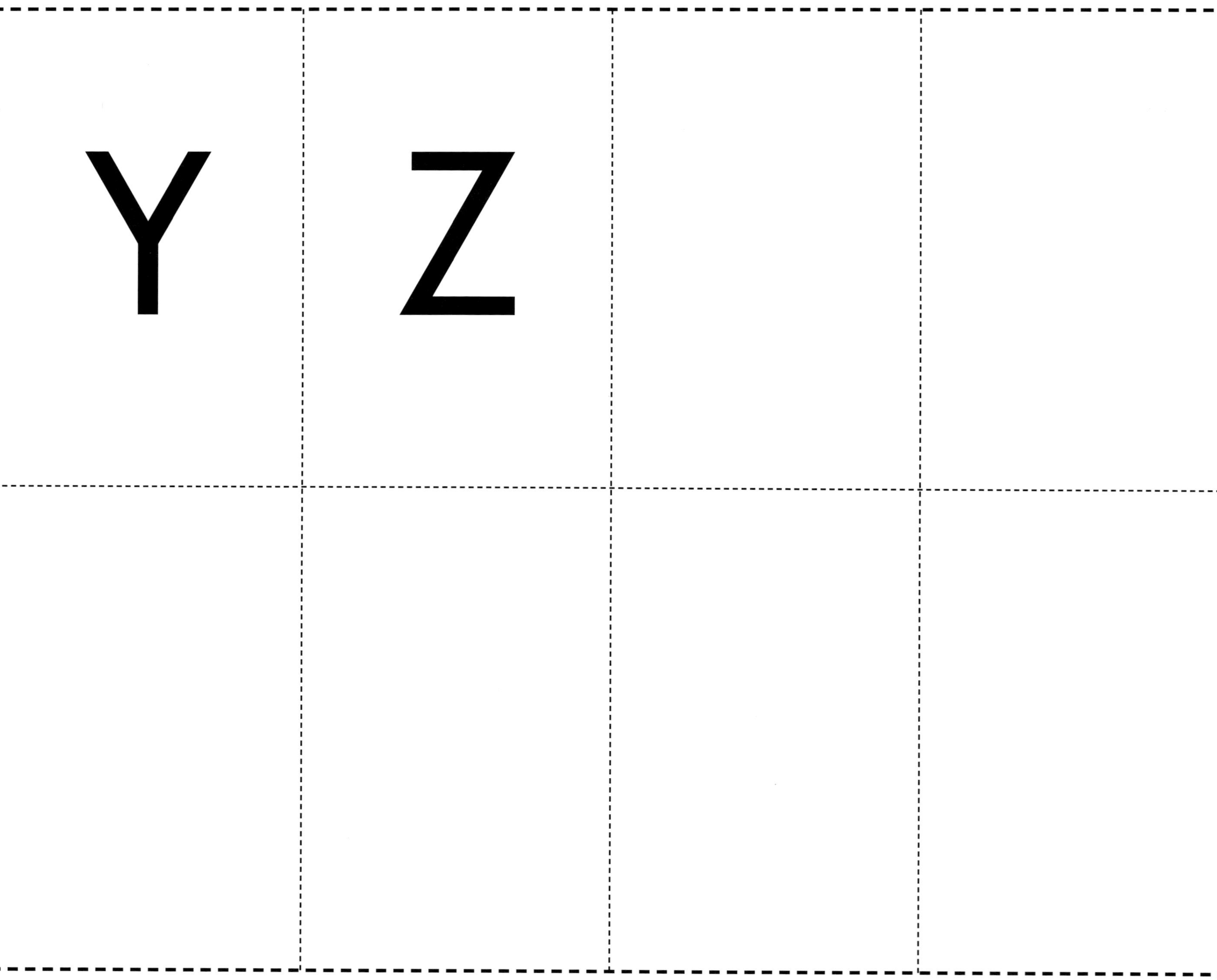

© 1999 LeapFrog SchoolHouse

N

K

© 1999 LeapFrog SchoolHouse

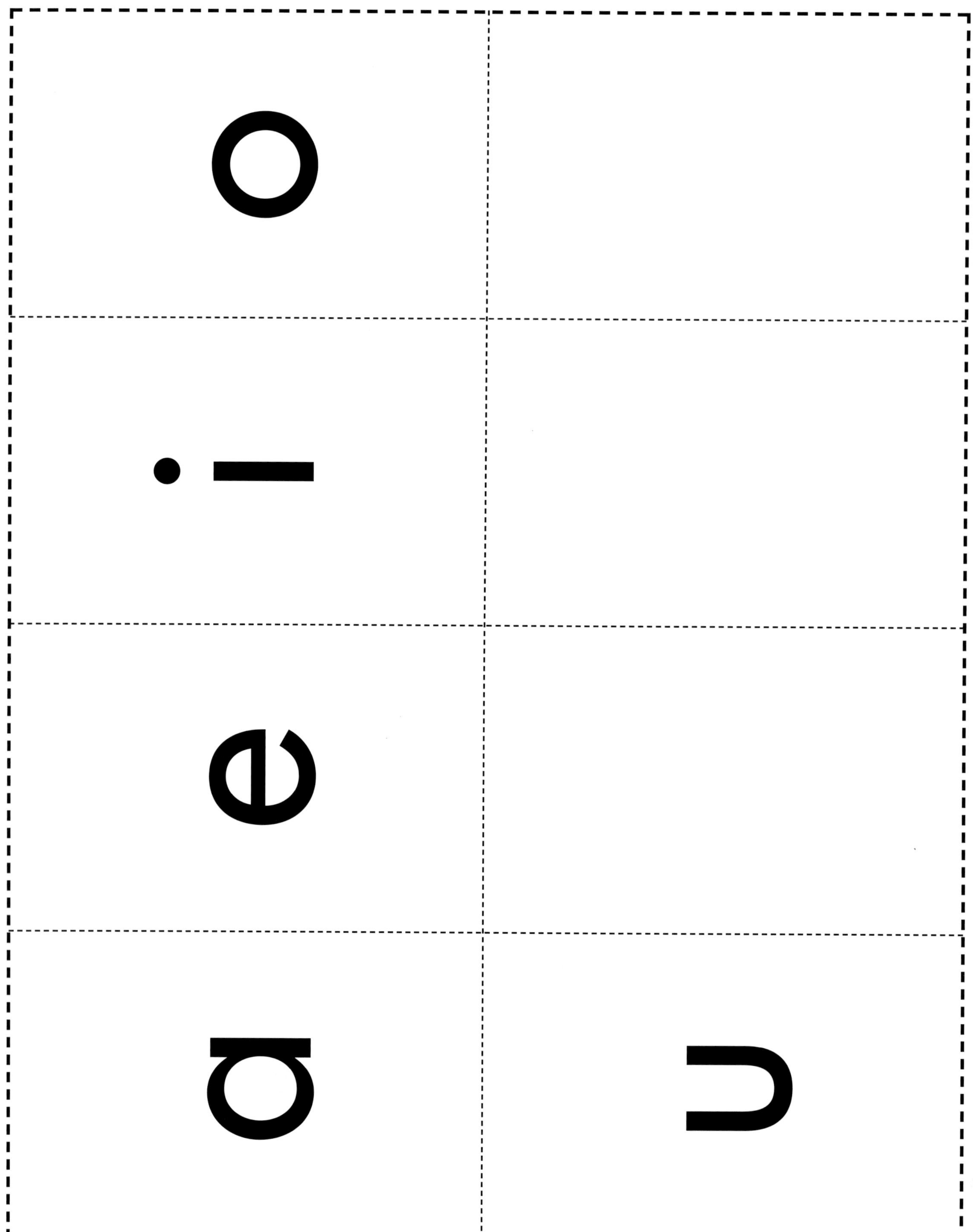

© 1999 LeapFrog SchoolHouse

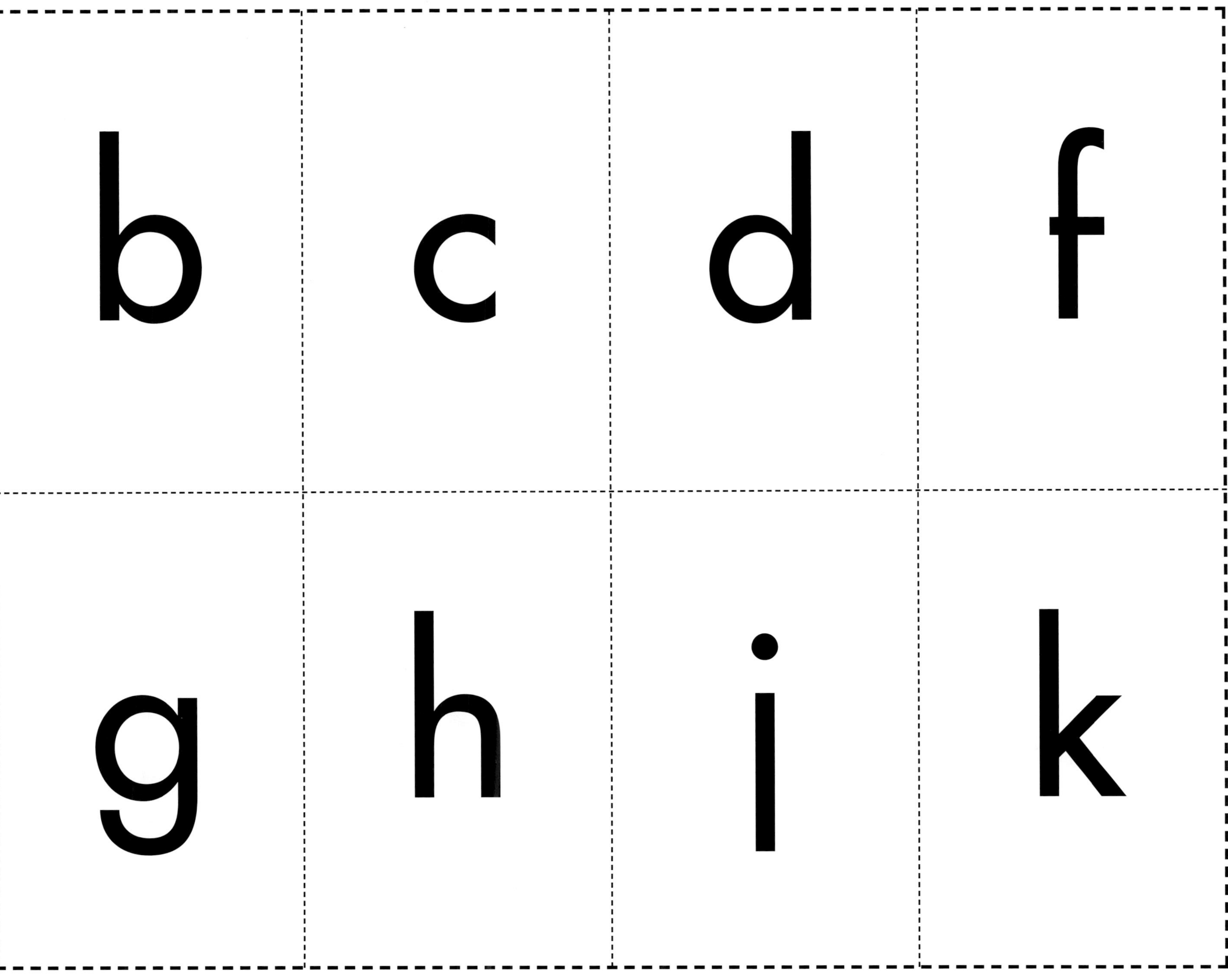

© 1999 LeapFrog SchoolHouse

p

t

n

s

m

r

l

b

© 1999 LeapFrog SchoolHouse

Extra Consonant Cards

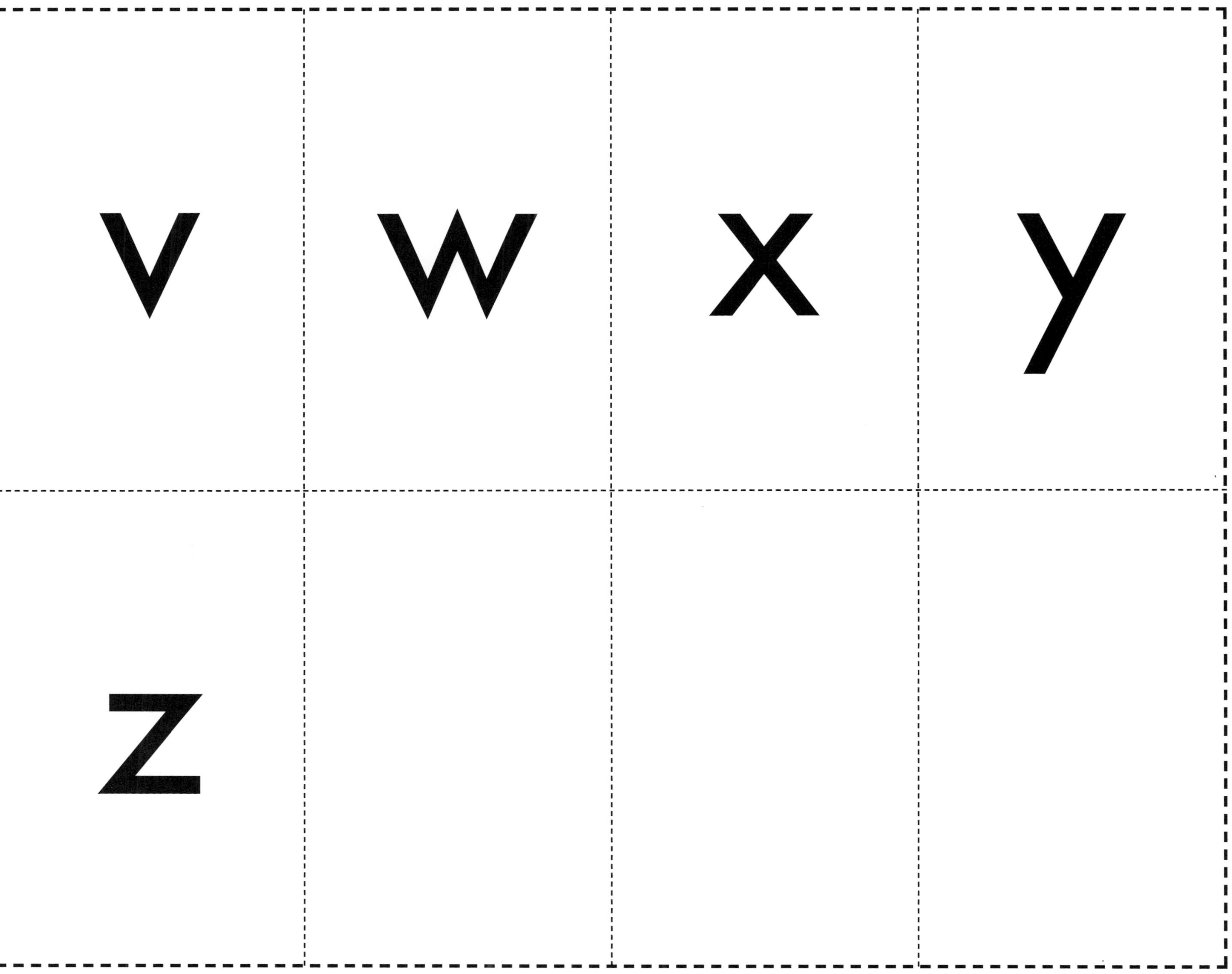

© 1999 LeapFrog SchoolHouse

Leap's
Big, Big Bag

Written by Rozanne Lanczak Williams
Illustrated by Ann Iosa

A Division of
Knowledge Kids Enterprises, Inc.

Leap has a lid.

Assembling your black line master.

1. Photocopy with page 7 backing up to page 8, and with page 9 backing up to page 10, (shown at right).

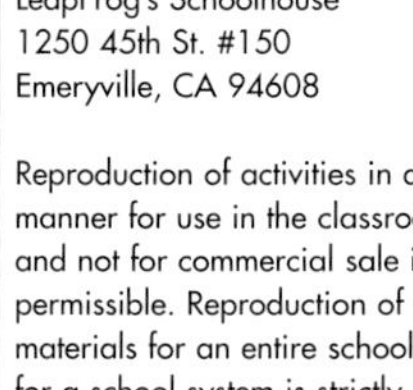

2. Align the sheets so that pages 8 and 10 are at top left, with page 10 on top.

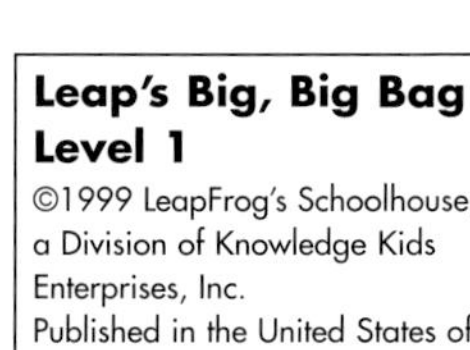

3. Fold the sheets down so that pages 6 and 7 face you. Cut along the fold.

4. Fold the sheets across to the right so that the cover faces you.

5. Open and staple on the fold.

**Leap's Big, Big Bag
Level 1**
©1999 LeapFrog's Schoolhouse, a Division of Knowledge Kids Enterprises, Inc.
Published in the United States of America by:
LeapFrog's Schoolhouse
1250 45th St. #150
Emeryville, CA 94608

Reproduction of activities in any manner for use in the classroom and not for commercial sale is permissible. Reproduction of these materials for an entire school or for a school system is strictly prohibited.

With a box
and a tag,
and a pot, lid
and rag,

Leap's Big, Big Bag
Level 1

Short vowel story
CVC (Consonant, Vowel, Consonant)

Sight Words	Challenging Words
with the look he do	things Leap

Discussion Ideas

1. Name some things that Leap used to make the robot.

2. What else could Leap make using the things in his bag?

3. What would you use to make a robot?

Leap has a tag.

With a rip
and a zip!

Leap has a box.

With a zig
and a zag!

Written by Rozanne Lanczak Williams
Illustrated by Ann Iosa

Leap's Big, Big Bag

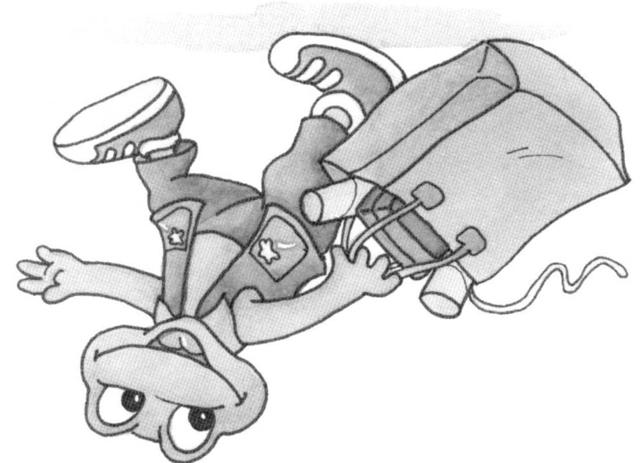

Look what Leap
did with the things
in his bag!

Leap has a rag
in his big, big bag.

what can he do
with the things
in his bag?

© 1999 LeapFrog SchoolHouse

Picture Cards

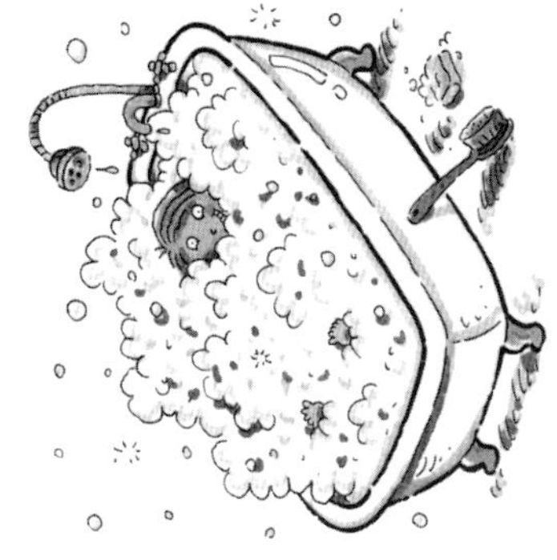

© 1999 LeapFrog SchoolHouse

© 1999 LeapFrog SchoolHouse

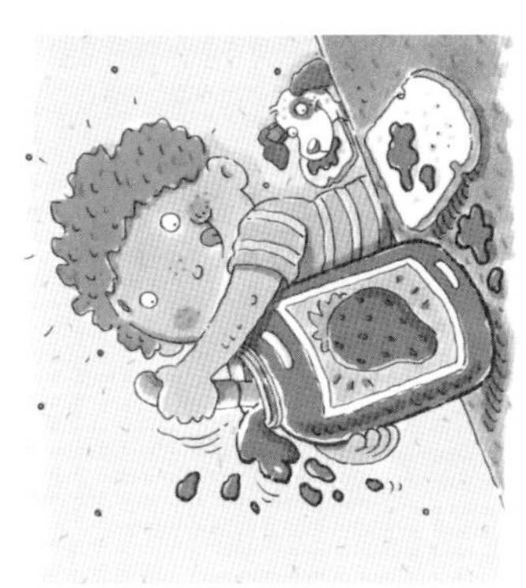

© 1999 LeapFrog SchoolHouse

BINGO

A	K	G	T
F	R	S	E
U	M	Y	I
W	P	H	Z

© 1999 LeapFrog SchoolHouse

© 1999 LeapFrog SchoolHouse

BINGO

B	L	N	E
T	O	F	R
X	D	J	I
S	G	M	Q

© 1999 LeapFrog SchoolHouse

© 1999 LeapFrog SchoolHouse

BINGO

M	A	L	B
F	O	S	U
W	N	Z	Q
X	C	V	K

© 1999 LeapFrog SchoolHouse

© 1999 LeapFrog SchoolHouse

BINGO

W	E	R	T
Y	U	I	P
A	S	D	F
V	B	N	M

© 1999 LeapFrog SchoolHouse

© 1999 LeapFrog SchoolHouse

bingo

n	g	x	p
r	h	m	e
a	w	t	f
b	i	v	s

© 1999 LeapFrog SchoolHouse

© 1999 LeapFrog SchoolHouse

bingo

l	k	j	h
s	d	f	g
a	c	w	e
x	p	m	n

© 1999 LeapFrog SchoolHouse

© 1999 LeapFrog SchoolHouse

bingo

l	e	v	t
o	s	d	n
z	v	r	y
a	h	u	r

© 1999 LeapFrog SchoolHouse

© 1999 LeapFrog SchoolHouse

bingo

m	f	x	o
q	i	l	k
b	p	r	d
v	u	c	a

© 1999 LeapFrog SchoolHouse

© 1999 LeapFrog SchoolHouse

bingo

© 1999 LeapFrog SchoolHouse

© 1999 LeapFrog SchoolHouse

ro_

co_

ri_

cu_

re_

ca_

ra_

ru_

© 1999 LeapFrog SchoolHouse

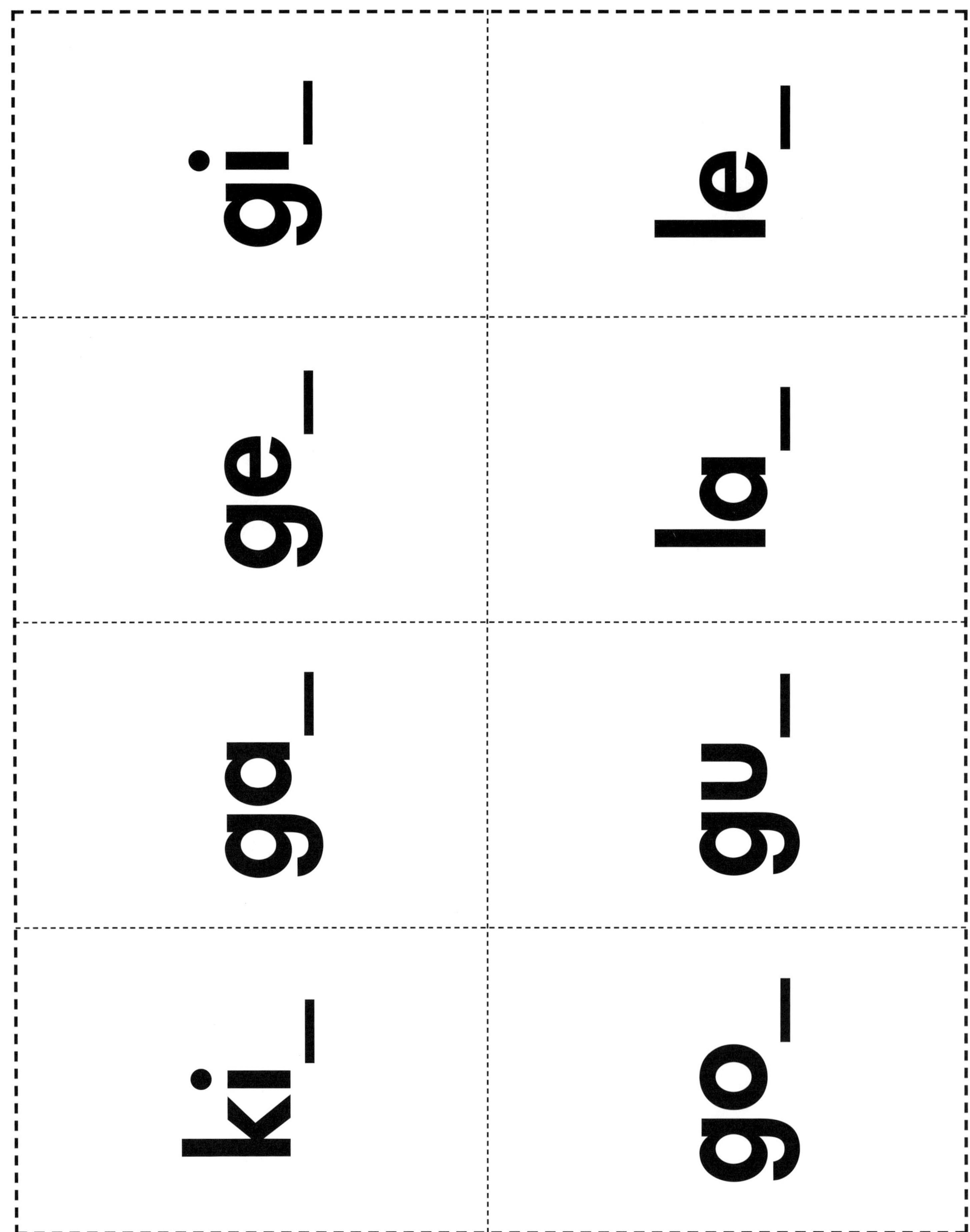

© 1999 LeapFrog SchoolHouse

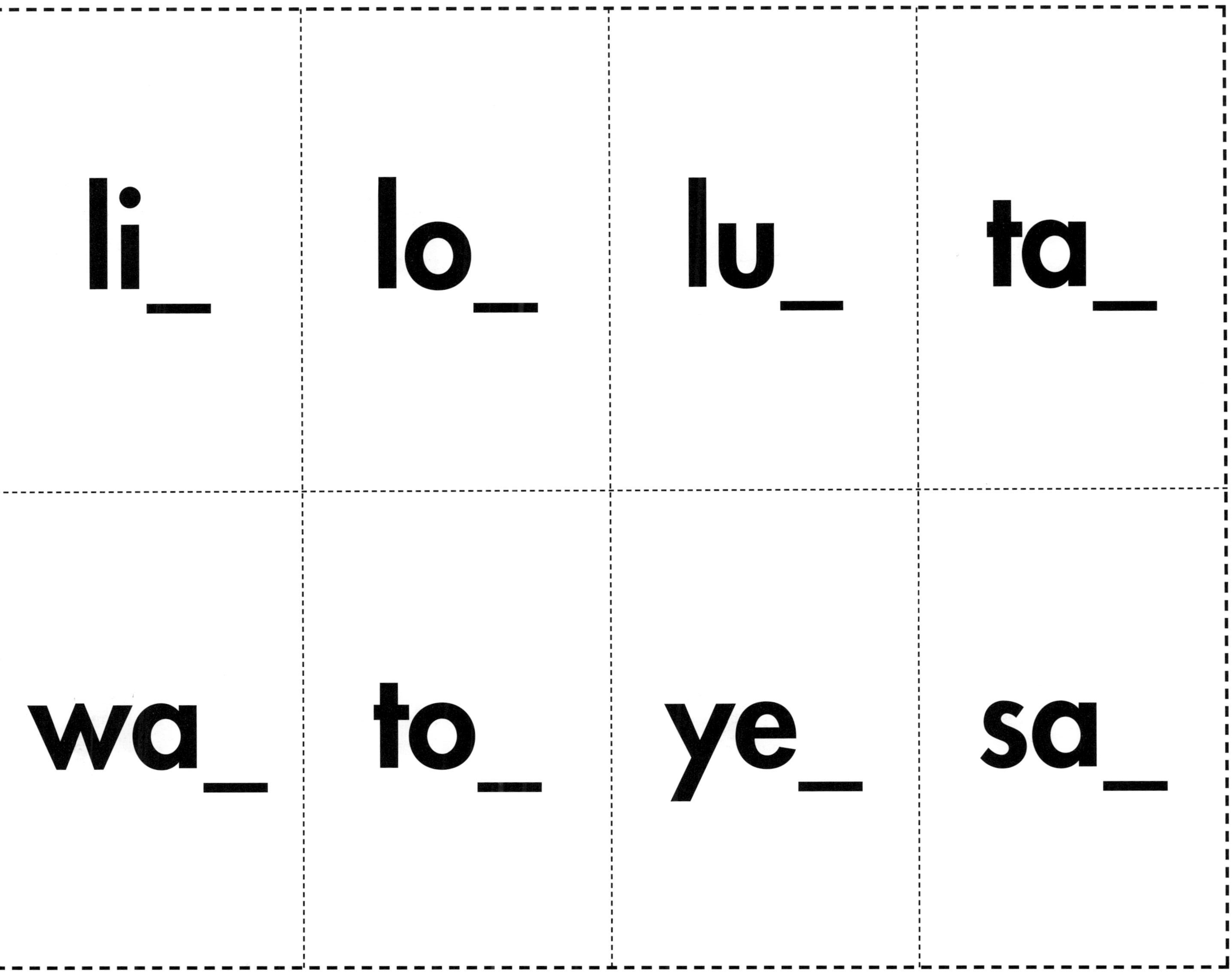

© 1999 LeapFrog SchoolHouse

su_

du_

so_

do_

si_

di_

se_

da_

© 1999 LeapFrog SchoolHouse

ho_

mo_

hi_

me_

he_

ma_

ha_

hu_

© 1999 LeapFrog SchoolHouse

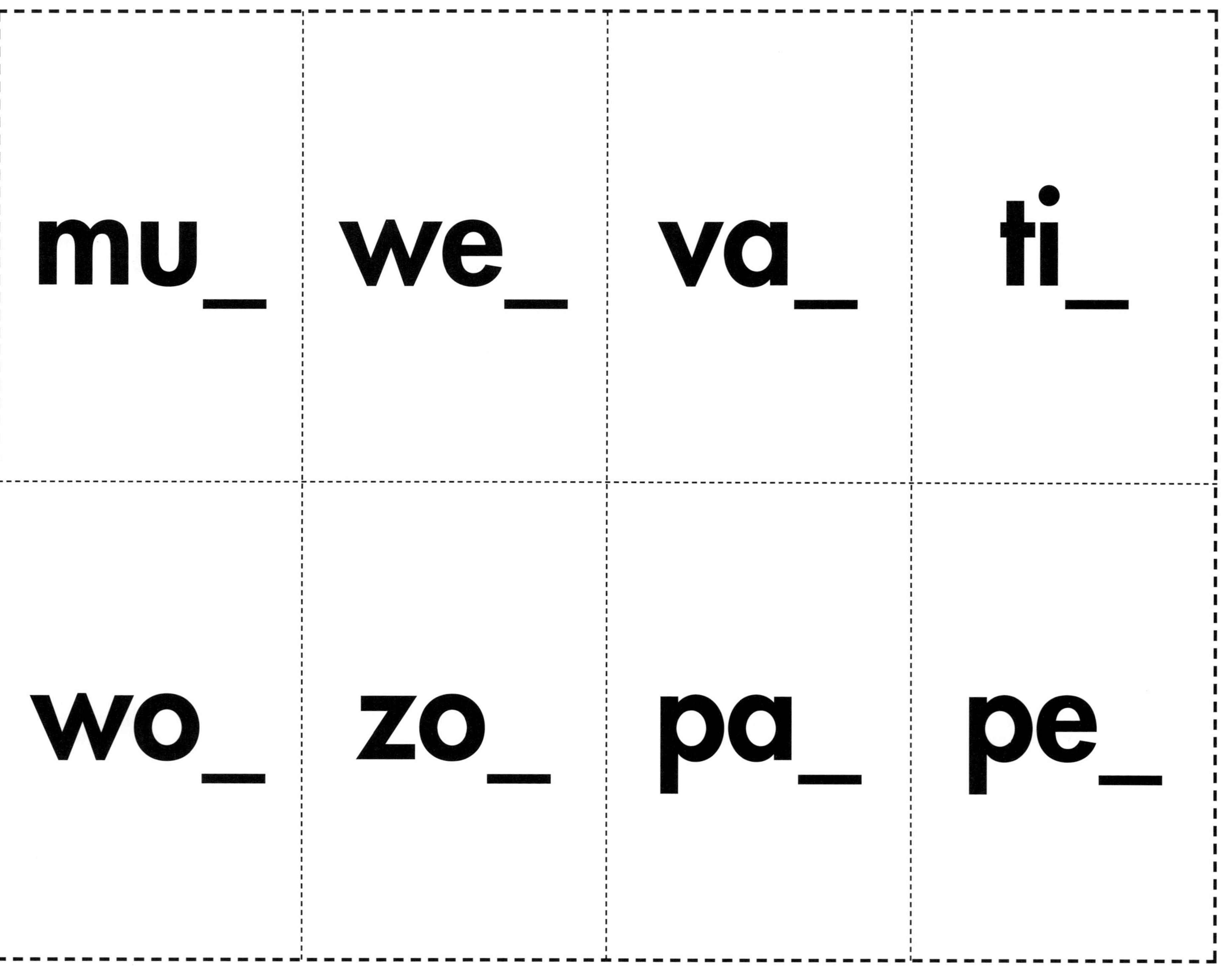

© 1999 LeapFrog SchoolHouse

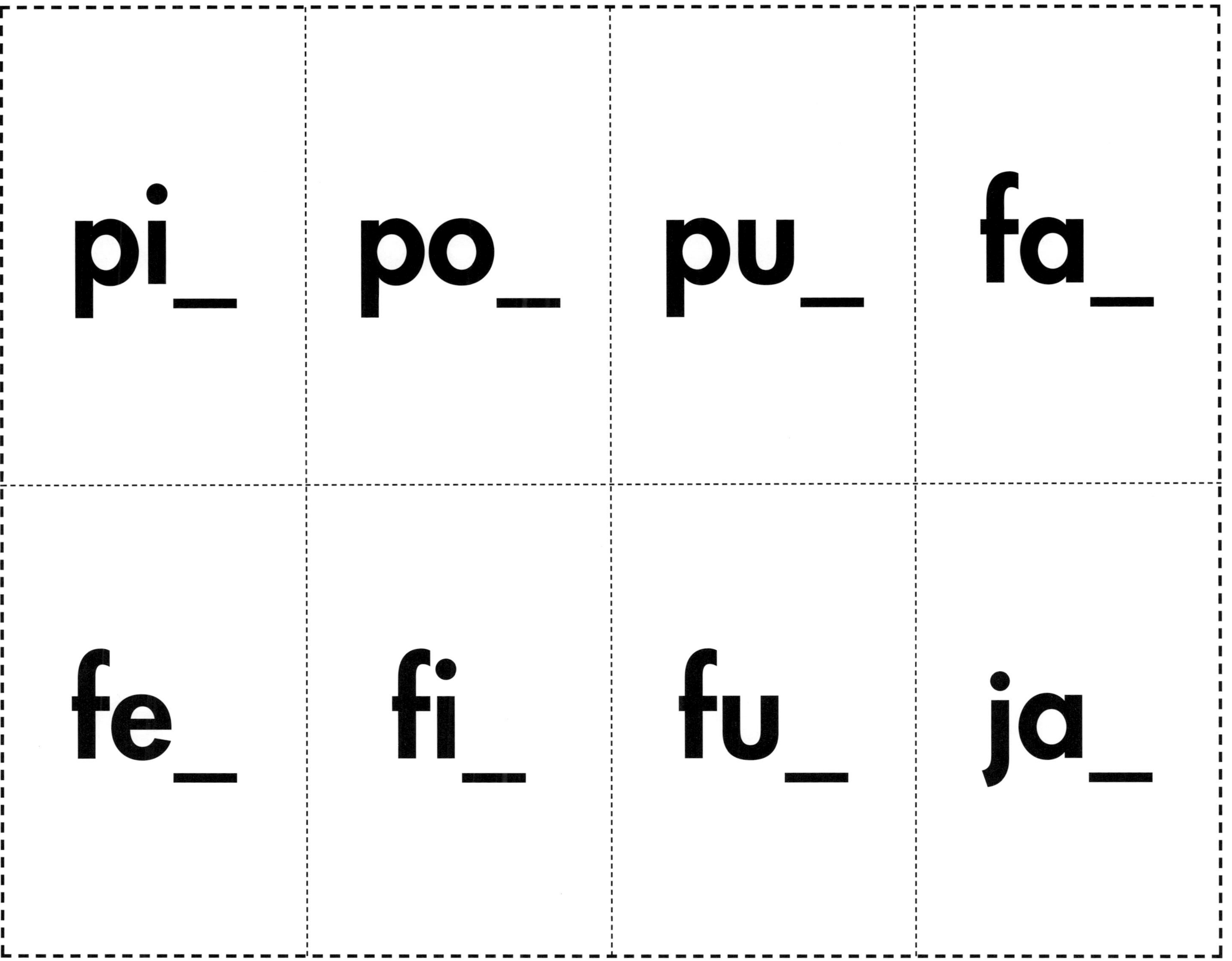

© 1999 LeapFrog SchoolHouse

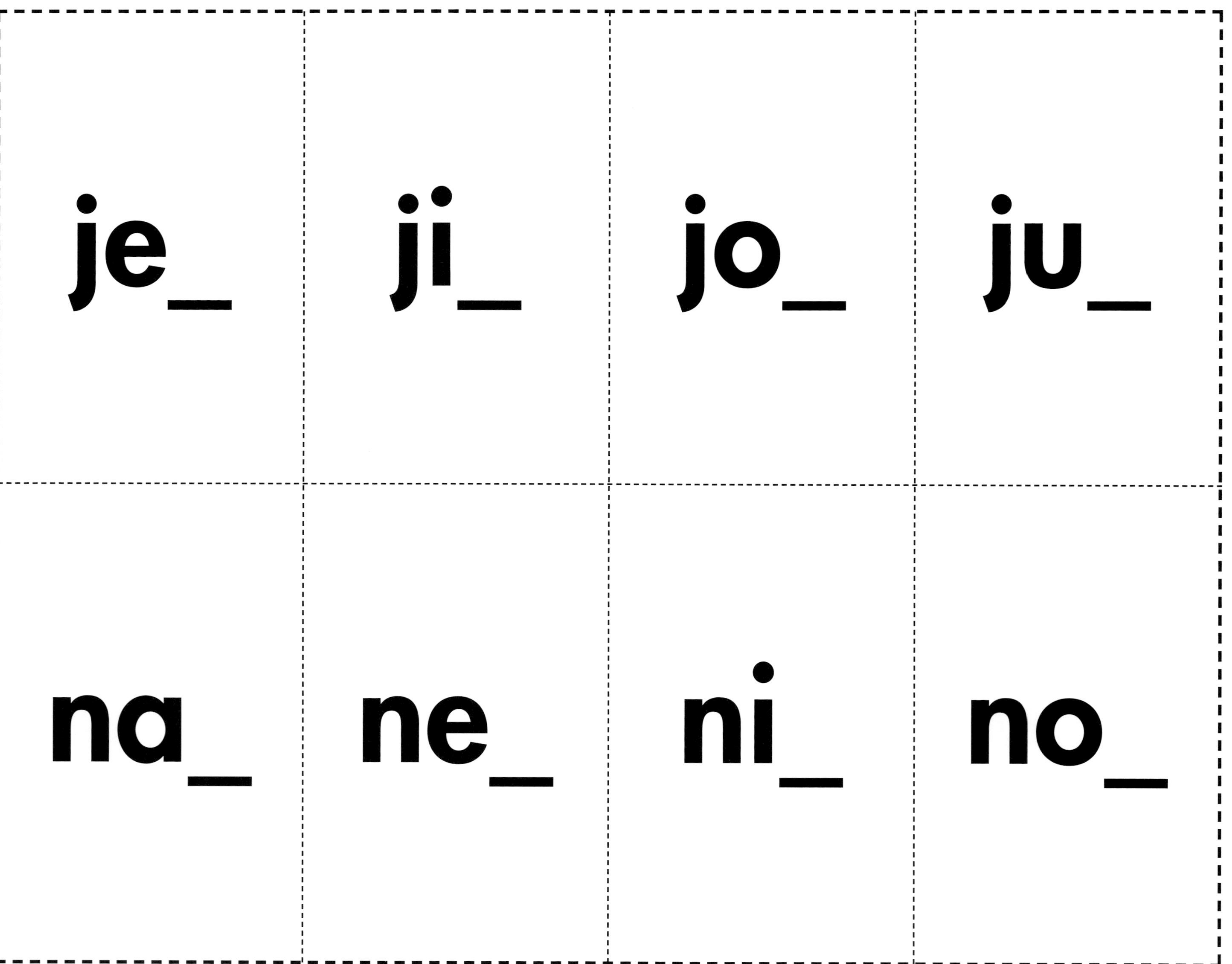

© 1999 LeapFrog SchoolHouse

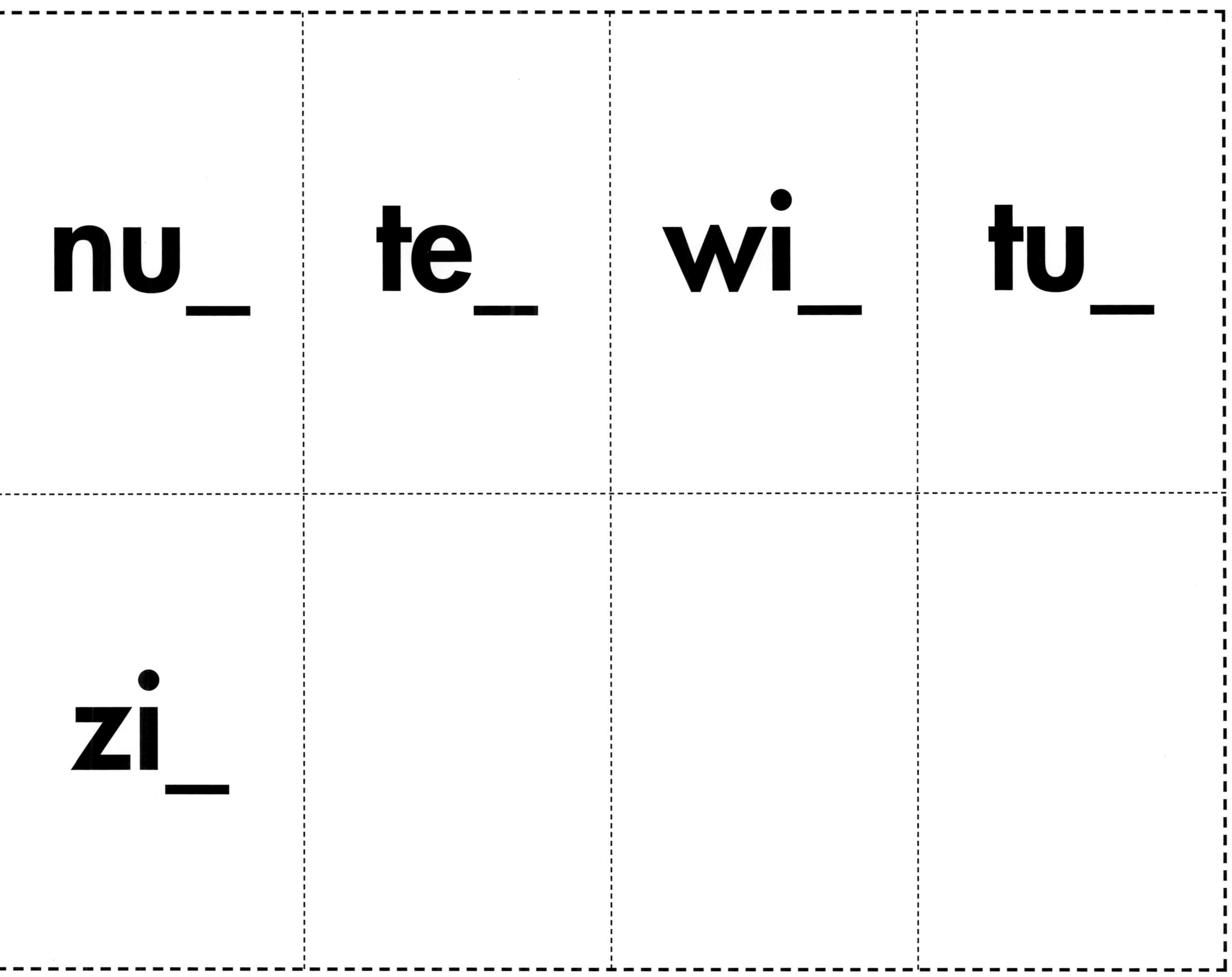

© 1999 LeapFrog SchoolHouse

_un

_en

_at

_am

_et

_an

_ug

_us

© 1999 LeapFrog SchoolHouse

Word Family Cards

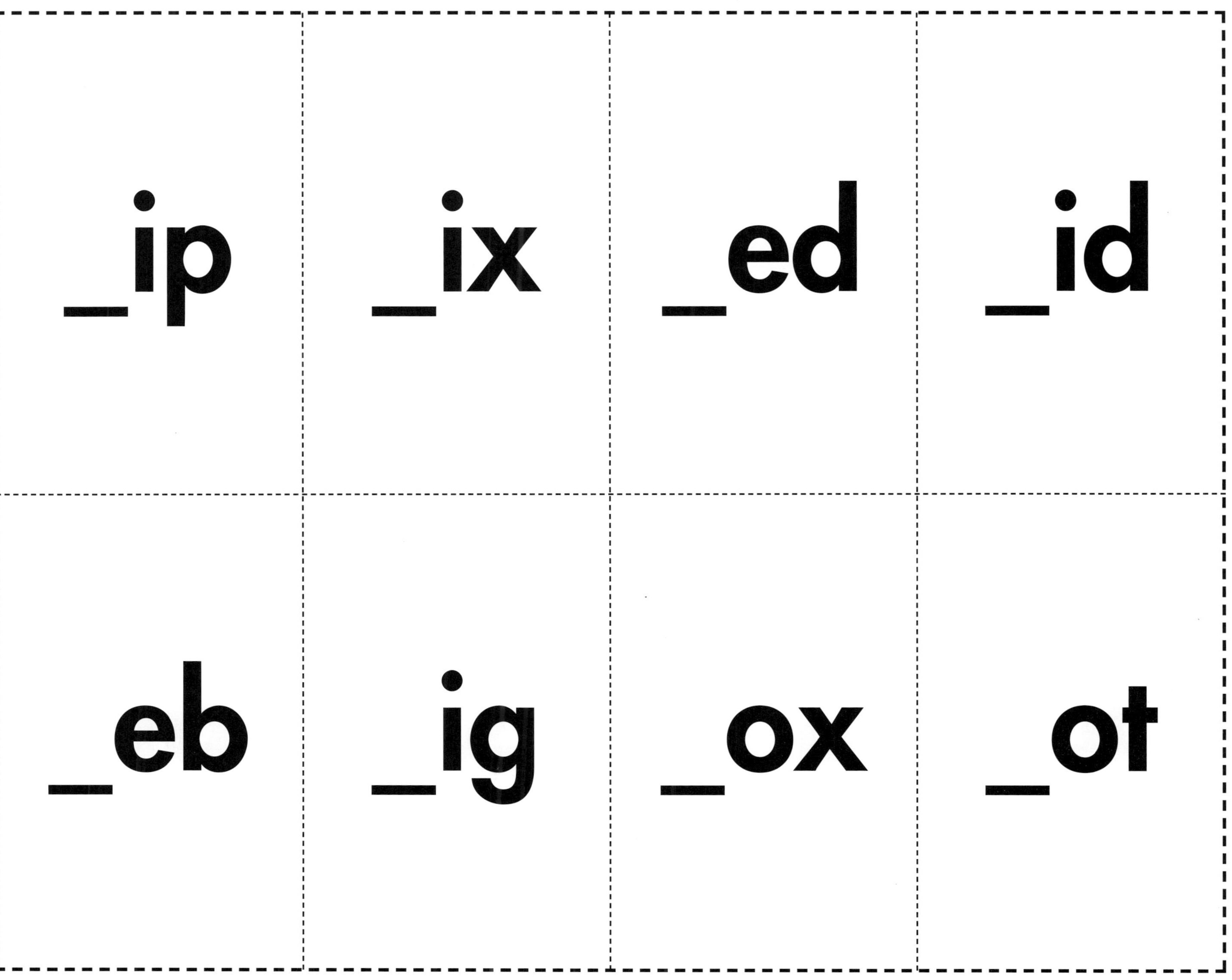

© 1999 LeapFrog SchoolHouse

_id _og _ub _op

© 1999 LeapFrog SchoolHouse

bingo

fan	kid	pot
box	mug	net
man	bus	top

© 1999 LeapFrog SchoolHouse

© 1999 LeapFrog SchoolHouse

bingo

sun	cat	fox
mug	net	kid
pan	bus	jam

© 1999 LeapFrog SchoolHouse

© 1999 LeapFrog SchoolHouse

bingo

jet	zip	bed
jam	rug	web
lid	cat	box

© 1999 LeapFrog SchoolHouse

© 1999 LeapFrog SchoolHouse

bingo

tub	van	jam
lid	man	bun
hen	mug	fan

© 1999 LeapFrog SchoolHouse

© 1999 LeapFrog SchoolHouse

bingo

rag	dog	bed
mix	pig	tub
fan	log	top

© 1999 LeapFrog SchoolHouse

© 1999 LeapFrog SchoolHouse

bingo

lid	bus	rag
dog	pan	pig
web	jet	tag

© 1999 LeapFrog SchoolHouse

© 1999 LeapFrog SchoolHouse

bingo

box	sun	pan
jet	fox	van
fin	net	mop

© 1999 LeapFrog SchoolHouse

© 1999 LeapFrog SchoolHouse

bingo

pot	bed	rug
van	pig	kid
mug	fan	top

© 1999 LeapFrog SchoolHouse

© 1999 LeapFrog SchoolHouse

bingo

© 1999 LeapFrog SchoolHouse

© 1999 LeapFrog SchoolHouse

CVC

CVC

CVC

CVC

CVC

CVC

CVC

CVC

CVC

© 1999 LeapFrog SchoolHouse

box	jig	mop	cap	top
lid	fun	zip	sat	pat
pig	pop	kid	hay	run
sun	men	cat	rag	rub
fox	leg	jam	rug	but
net	dad	dog	top	bag
pan	can	joy	tub	bad
van	sit	tap	bun	cab
hen	lay	rig	bed	hat
jet	tag	toy	fan	set
fin	pot	pup	bus	jay
yes	mug	big	mix	
can	man	lag	yet	
hop	web	dig	boy	

© 1999 LeapFrog SchoolHouse

sun

van

pig

pan

lid

net

box

fox

© 1999 LeapFrog SchoolHouse

tag

web

fin

man

jet

mug

hen

pot

© 1999 LeapFrog SchoolHouse

cat

rug

kid

rag

zip

dog

mop

jam

© 1999 LeapFrog SchoolHouse

bed	mix
bun	bus
tub	log
top	fan

© 1999 LeapFrog SchoolHouse

Picture Cards (no words)

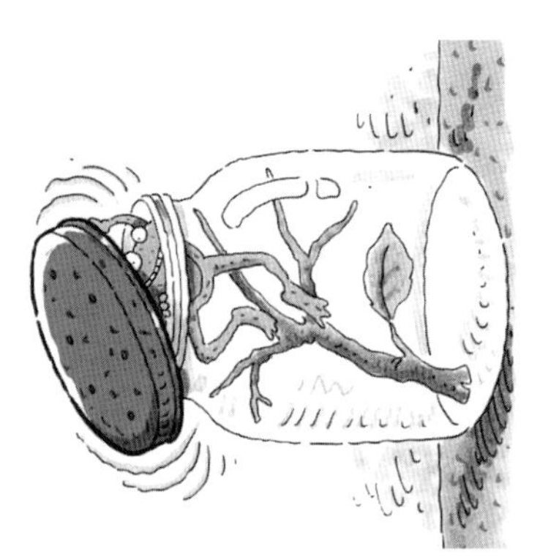

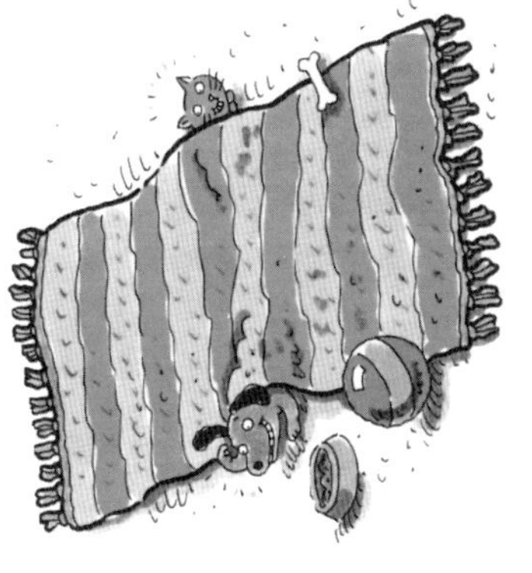

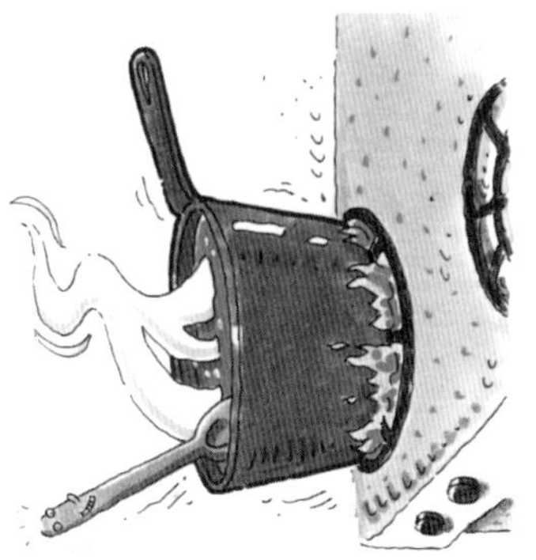

© 1999 LeapFrog SchoolHouse

Picture Cards (no words)

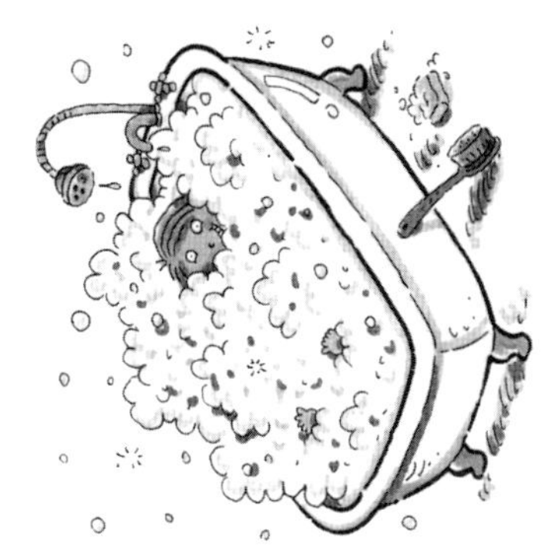

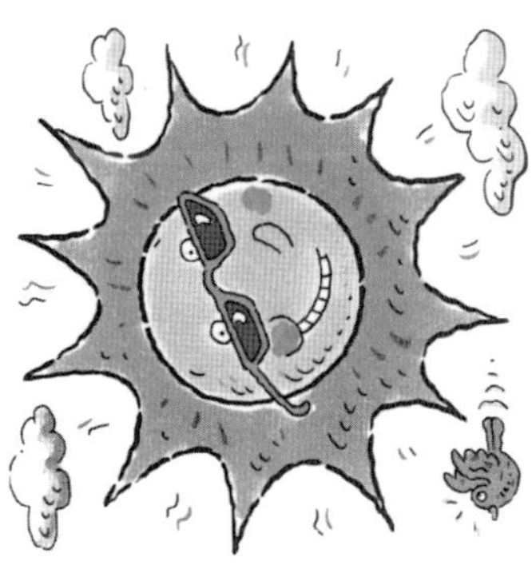

© 1999 LeapFrog SchoolHouse

Picture Cards (no words)

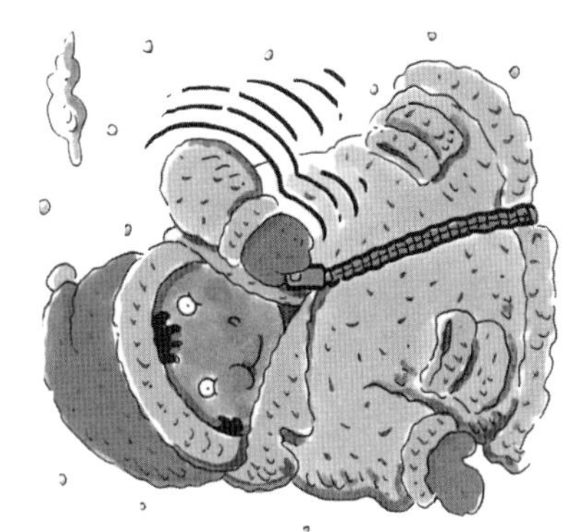

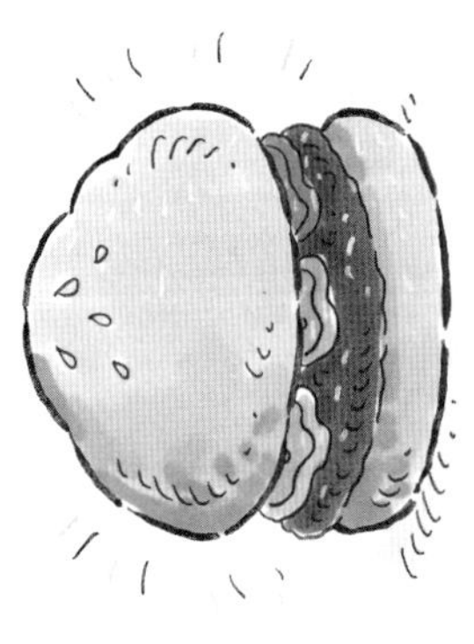

© 1999 LeapFrog SchoolHouse

Picture Cards (no words)

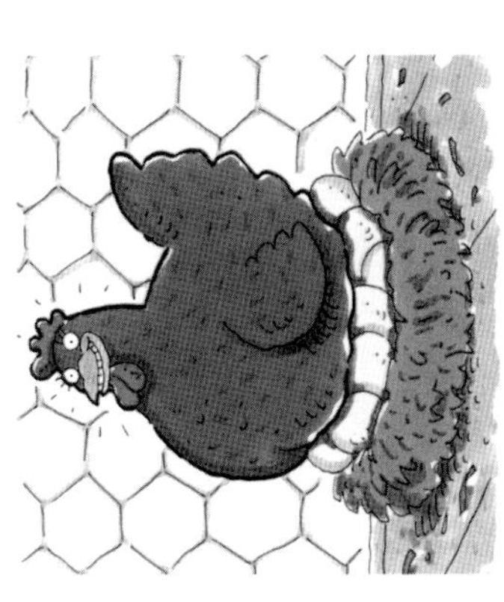

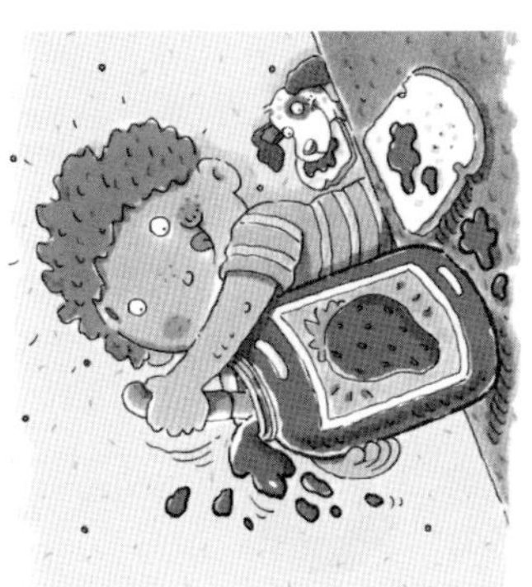

© 1999 LeapFrog SchoolHouse

i d l

u g m

a g r

i g p

a g t

u g r

o b x

o p t

© 1999 LeapFrog SchoolHouse

o d g

e n t

a m n

u b t

u n s

e b w

o p t

o f x

© 1999 LeapFrog SchoolHouse

i p z

a f n

a n p

i d k

u b n

a n v

o m p

e b d

© 1999 LeapFrog SchoolHouse

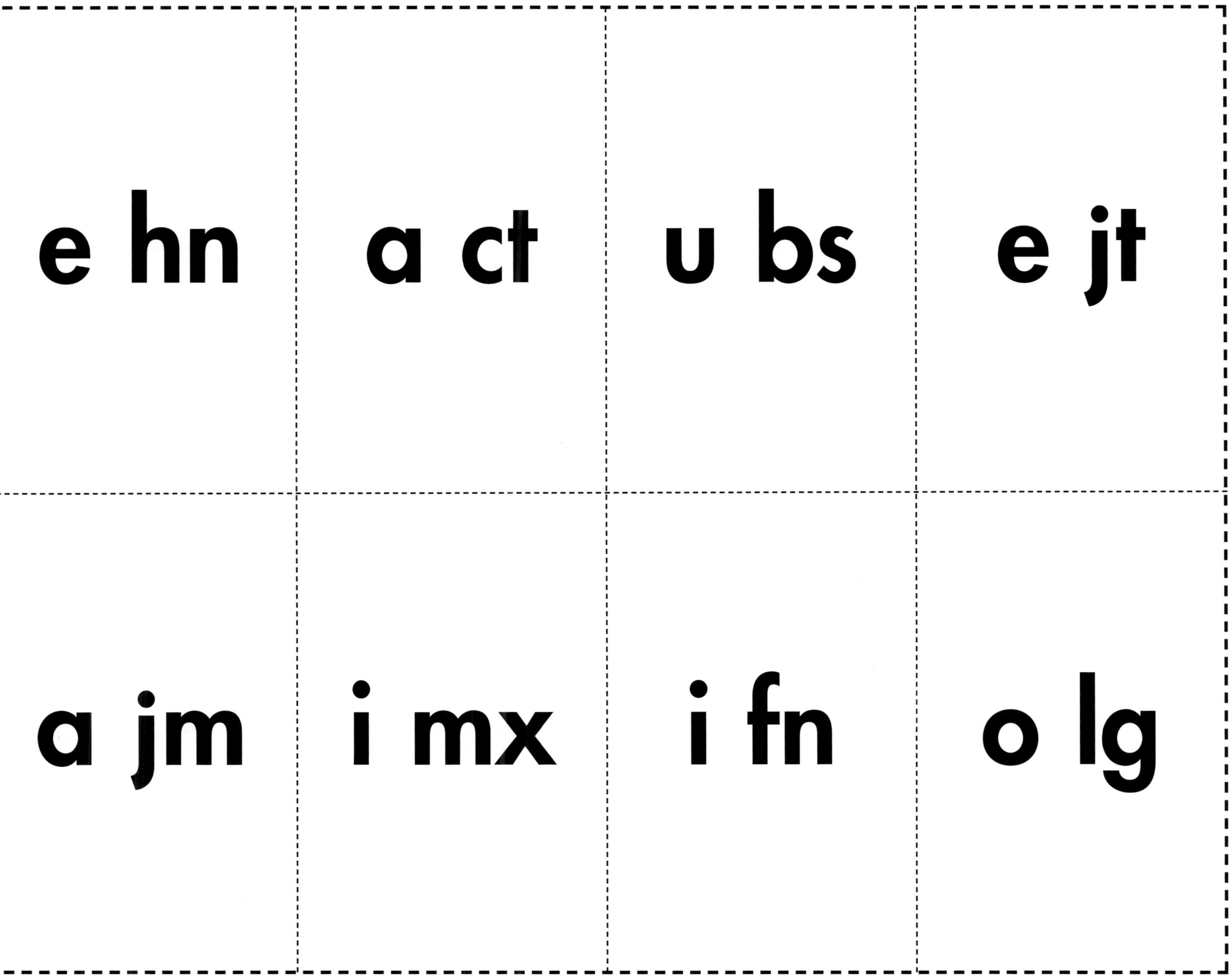

© 1999 LeapFrog SchoolHouse

g	n	u	
f	m	t	
e	l	s	z
d	k	r	y
c	j	q	x
b	i	p	w
a	h	o	v

© 1999 LeapFrog SchoolHouse

LeapMat™ Activity Matrix

Activities Completed (note date completed)

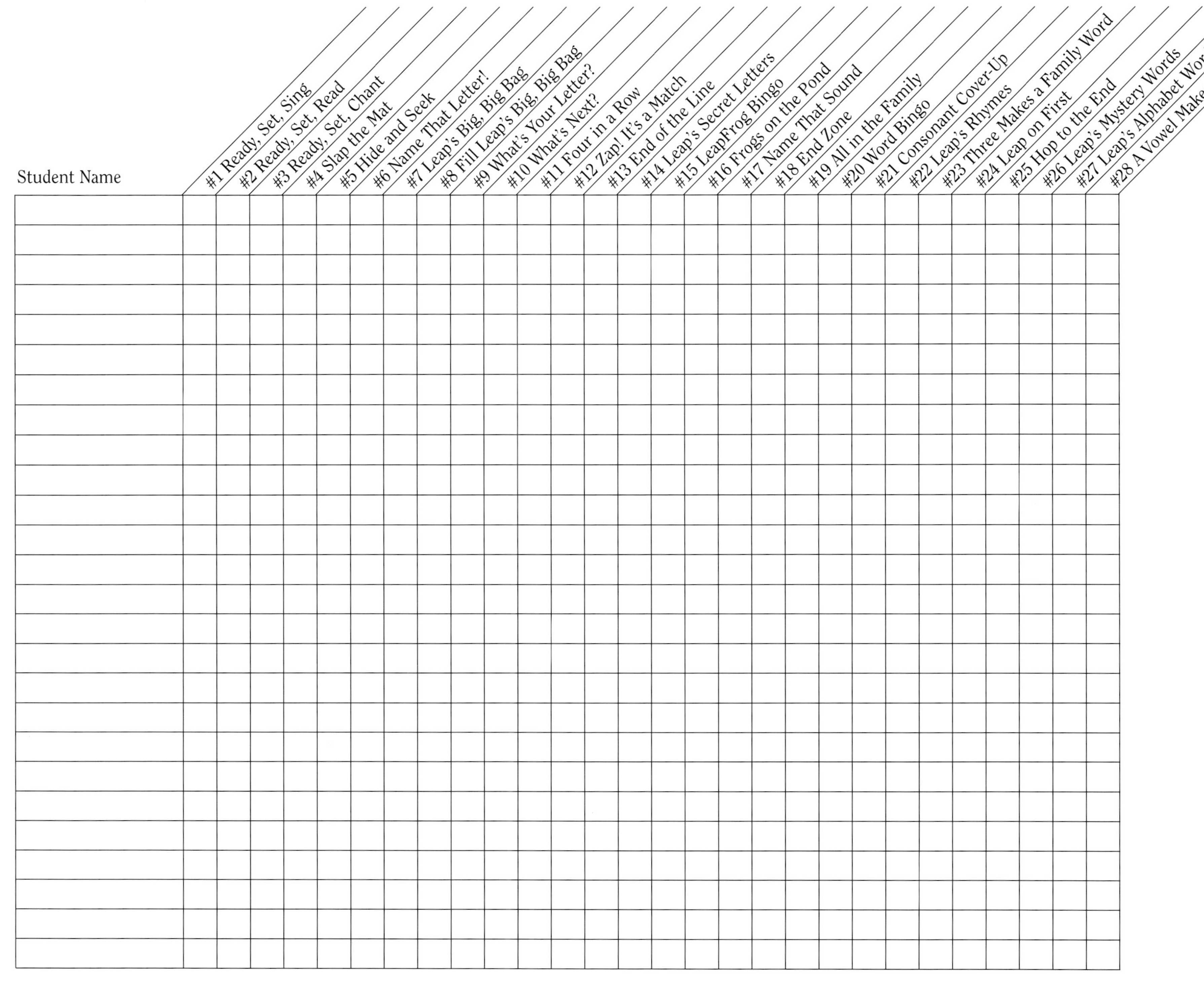

© 1999 LeapFrog SchoolHouse

Caring for the LeapMat

To ensure continued enjoyment of your LeapMat:

- Keep away from food and beverages and avoid spills.
- Do not submerge the LeapMat in water. Use a slightly damp cloth to keep the LeapMat clean and free from dirt.
- Remove batteries for prolonged storage
- Avoid extreme hot or cold temperatures
- The LeapMat is not to be used on slippery surfaces or for sliding
- Non-rechargeable batteries are not to be recharged.

Battery Safety

Follow these guidelines when installing batteries:

- Adults should always install batteries
- Follow the polarity (+/-) carefully.
- Do not mix old and new batteries or types of batteries (i.e. alkaline, standard, rechargeable). Do not use rechargeable batteries.
- Dispose of used batteries properly.
- Do not incinerate used batteries.

©1999 LeapFrog SchoolHouse